BARN CONVERSIONS

37 INSPIRATIONAL PROJECTS

304 PAGES ■ 301 COLOUR PHOTOS ■ 49 COLOUR ILLUSTRATIONS

BARN CONVERSIONS

SECOND EDITION

Ovolo Publishing
1 The Granary, Brook Farm,
Ellington, Huntingdon,
Cambridgeshire
PE28 0AE

ISBN: 978 0 954867485

All of the material in this book has previously appeared in Homebuilding & Renovating magazine - Britain's best selling monthly for self-builders and renovators (www.homebuilding.co.uk). Some of the homes featured in this edition of the book previously appeared in the first edition.

Book Design: Gill Lockhart

This edition first published in the UK by Ovolo Publishing Ltd
Printed in Italy

For more information on books about property and home interest please:
visit: www.ovolopublishing.co.uk
email: info@ovolopublishing.co.uk
or call: 01480 891595 (24 hours)

CONTENTS

100
162
204
50
224
186

240
250
258
268
288
276

FOREWORD

ON A BACK road there is a brick barn that I drive past regularly. Over the years I have watched it disintegrate as nature and vandals reduced it from a proud, watertight structure to a ruin with gaping holes in the roof and jagged openings where doors and window frames have rotted or been stolen. It's crying out be to brought back into useful life. But the farmer doesn't want to sell it. "We never sell anything," he says proudly. So the barn rots.

Why am I telling you this? Because there are still those who say that barns should not be converted. That barns should be for farms – not people. Many planners subscribe to this view – seeing conversion to residential use as only marginally better than demolition! But that is not reality. These buildings are redundant. And this book proves that there is life after agriculture for these structures.

With sympathetic conversion, barns and other former agricultural buildings can make a spectacular contribution to the architectural diversity of our countryside. The aim, in most cases, is to retain the exterior appearance as if it is still a working building . And let's face it, there is little point in paying for a building as individual as a barn and turning into something that looks like any other house.

And it is the desire to create something individual – yet reflecting the building's past – that is the common thread running through the 37 case studies in this second edition of *The Homebuilding & Renovating Book of Barn Conversions.* Thankfully, for those want a barn to convert, there are still plenty of farm buildings awaiting restoration. As these case studies show, they range from the humble to the grand – but all are capable of being transformed into homes which offer something special.

The big threshing barns may be the cathedrals of the countryside, with soaring vaulted ceilings which offer the possibility of double and triple height rooms, but smaller buildings can offer an intimacy which is also appealing. And almost all are in great locations.

Whether you are looking for traditional-style or contemporary interiors – where chrome and steel contrast with the mellow tones of the original structure – this book is for you!

'WITH SYMPATHETIC CONVERSION, BARNS CAN MAKE A SPECTACULAR CONTRIBUTION TO THE ARCHITECTURAL DIVERSITY OF OUR COUNTRYSIDE'

Michael Holmes, Editor-in-Chief, Homebuilding & Renovating magazine

A LOW COST DIY CONVERSION OF AN OLD BARN

ALL THEIR OWN WORK

Doug and Suzy Palmer have spent eight years and just £135,000 converting their Victorian brick and flint barn, using reclaimed materials and tackling all of the work themselves.

WORDS: DEBBIE JEFFERY PHOTOGRAPHY: NIGEL RIGDEN

WHEN I FIRST saw our farmhouse in Hampshire 20 years ago it was the complex of brick and flint outbuildings which attracted my attention," says Doug Palmer. "I felt sure that, one day, these would be easy to convert into a fantastic home. Little did I realise that this would turn into a DIY job of gigantic proportions. I didn't dare say anything to Suzy – I saved that for the day after we moved in!"

The stresses and strains of running a successful restaurant business, together with bringing up a young family, meant that the couple waited for some years before commencing the conversion project. During this time they renovated and extended the farmhouse, landscaping much of the surrounding five acres. The site had previously been used for car repairs and as a small petrol station and transport café, leaving the whole area littered with rusting remains of vehicles, old petrol pumps and piles of scrap.

Although the Palmers had given much thought to the eventual layout and design for the barn, finding an architect to interpret their ideas and come up with a suitable scheme proved taxing.

"There were several factors which had to be reconciled," says Doug. "Suzy was insistent that the kitchen should have the same southerly view down the valley that we already enjoyed from the farmhouse and, as the barn faces east-west, this seemed somewhat difficult. Secondly, much of the main barn was not high enough to accommodate two floors and, lastly, it appeared that in order to raise the insulation levels to the required standard we would have to dry-line the existing walls – which would have taken away the very solid and rustic feel we so wanted to keep."

Ultimately, the problem of the kitchen was solved by adding a single storey pitched

"THE WALLS DIDN'T GO DOWN BELOW GROUND LEVEL, WE WOULD HAVE TO SIT THE ENTIRE BARN ON A RING OF CONCRETE..." ➤

The farmhouse style kitchen is housed in a new annexe added at the rear of the barn. The cupboards are built in reclaimed brick with simple ledged doors made from reclaimed timber.

Extra height was achieved by digging out one meter of the existing floor.

roof extension at right angles to the main barn, which not only provides wonderful southerly views but also encloses a sheltered terrace area and gives added width to the interior of the property. The dilemma of retaining a solid 'original' feel to the outer walls was resolved by building an inner skin of insulated reclaimed brick and block.

Structural engineer and family friend, Ian Calder, came up with a solution to the lack of headroom in the main barn. "Ian told us that we could get the extra height we needed by digging out about one metre of the existing floor. The bad news was that the walls didn't go down below ground level, and we would have to sit the entire barn on a ring of concrete extending down nearly two metres into the ground!" explains Doug, who by this time had sold his ➤

"THERE WERE TIMES WHEN WE DISAGREED ABOUT HOW TO DO SOMETHING AND IT WOULD END UP LIKE A SCENE FROM LAUREL AND HARDY..."

restaurant business, providing funding and giving him the time to work on the build.

A local bricklayer was employed on a labour-only basis to construct the new kitchen extension and build the chimney in the sitting room. Electricians and plumbers were hired when they were needed, and retired train driver, Cyril Gardner – who had already built his own house in a nearby village – was recruited to help with all the general building tasks. Together, Doug and Cyril tackled every aspect of the build including foundations, drainage, brick and flintwork, carpentry, roofing, and plastering.

The garden room, complete with original Victorian brick floor, links the main house to the guest suite.

"There were times when we disagreed about how to do something and it would end up like a scene from Laurel and Hardy," laughs Doug. "On the other hand, Cyril's help was invaluable and it would have been very difficult to work entirely alone. Having someone arrive on site each day at 8am made me stick to the task and put in full days."

Once the underpinning was complete, the internal floor was excavated, drainage laid and a new concrete floor poured. An insulated internal skin, constructed from reclaimed bricks and insulation blocks, was then built inside the original perimeter wall. Most of these original walls are constructed from chalk blocks with brick quoins and reveals, and Doug ➤

WILLS'S
STAR
STAR
Cigarettes

has replicated this in the new inner shell. Steel joists were then heaved into position – spanning the full width and running the entire length of the building. These pass over the inner skin wall and are supported by the outer walls, and Doug has inserted reclaimed pine beams under each one.

The family eventually moved into the barn at the end of 1999, five years after the start of building. Landscaping followed and was a massive undertaking. "As the ground floor was now nearly a metre below the outside ground level this had to be addressed all around the property," says Doug. "Fortunately I had invested in a compact tractor, with a front loader and mini digger attachment, and with this I was able to excavate, level and re-grade the land before building retaining walls and steps where necessary."

Doug then turned his attention to converting an attached stable building into a guest suite, which necessitated rebuilding two of the walls and replacing the entire roof. This area also provides a garden room, which links the main house to the guest suite. "We particularly wanted to keep this part of the building in its original state," says Suzy. "It's the only section that still had its Victorian brick floor so, by making it into a garden room, we avoided the need to change it in any way."

Measuring up to a metre in length by half a metre wide, and believed to have previously been the floor of a Lancashire cotton mill, the reclaimed slate flagstones weighed well over fifty kilos and were only moveable on a sturdy sack trolley. All were laid by Doug.

"I would definitely tackle any future building programme in a different way," says Doug. "I don't think I could face another eight year project! When I started this I was under the impression that 'self-build' meant that you had to build it yourself, but now I've heard of something called 'delegation' and I'm sure that's the way to go!" ■

FLOORPLAN

Three large double bedrooms, each with its own full size bathroom, are housed in the main barn with a guest suite in the attached stable conversion. A second, 'secret', staircase has been constructed behind the chimney in the sitting room to avoid dividing the barn along its length. This gives independent access to the master bedroom suite, and the lack of corridors maximises living space.

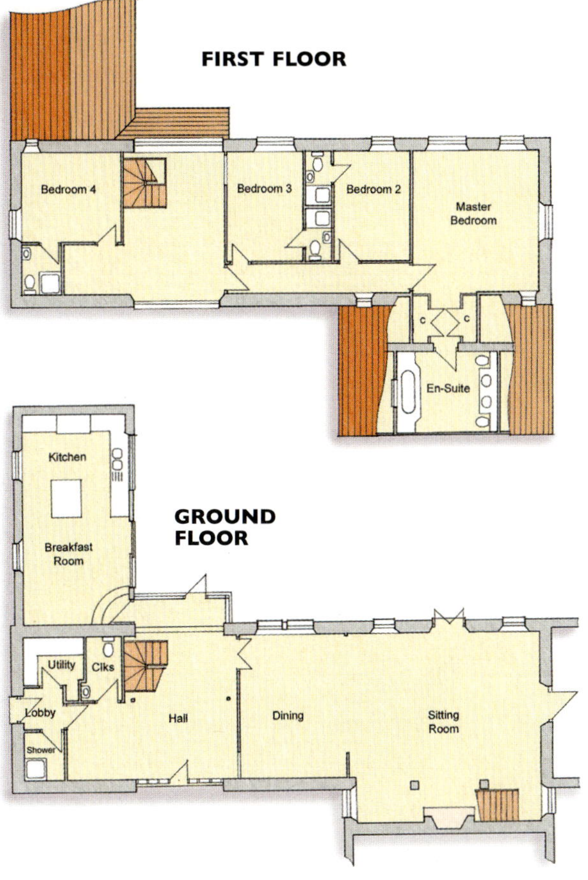

USEFUL CONTACTS

Architect Richard Ashby: 01420 538316; **Architectural technician** Robert Laverton: 01730 895316; **Structural engineer** Ian Calder: 01795 84334; **Plumbing & electrical supplies** Ideal Plumbing: 01420 88500; **Range cooker/boiler** Alpha Cookers Ltd.: 01635 582068; **Insulation** Seconds & Co: 01544 260501; **Joinery** Holmes & Pearcy: 01256 782575; **Underfloor heating** Eurogage Ltd.: 01342 323641; **Glass and windows** Darby Glass: 02392 69952; **Handmade tiles** Froyle Tiles: 01420 23693; **Reclamation yards** Walcot Reclamation: 01225 444404, Comley Lumber Centre: 01252 716882, Arthur Rudd Supplies: 01428 9814379, Brook Barn Specialists: 0118 9814379, Romsey Reclamation: 01794 524174, Drummonds 01428 609444

FACT FILE

Names: Doug and Suzy Palmer
Professions: Retired restaurateur and teacher
Area: Hampshire
House type: Four bedroom converted barn
House size: 415m²
Build route: DIY and subcontractors
Construction: Blockwork, brick and flint
Finance: Private
Build time: May '95 – May '03
Land value: £100,000 (already owned)
Build cost: £135,000
Total cost: £235,000
House value: £900,000
Cost/m²: £325

74%
COST SAVING

Cost Breakdown:

Fees and planning	£5,500
Underpinning	£8,183
Steel joists	£1,870
Double glazed units	£3,990
Alpha range cooker/boiler	£5,370
Reclaimed slate flagstone floorings	£10,575
Insulation	£2,333
Low voltage light fittings	£1,104
Window and door joinery	£7,448
Underfloor heating	£3,500
Wall and floor tiles	£2,227
Reclaimed bricks, timber and fittings	£5,500
Electric supply	£2,439
Builders' merchant	£39,000
Plumbing installation	£4,621
Sanitaryware and fittings	£18,315
Labour	£31,250
Subtotal	£153,225
Less VAT reclaim	–£18,350
TOTAL	**£134,875**

JUDGES SUMMARY

This well thought out barn conversion avoids many of the problems usually associated with adapting such buildings. The design, by architect Richard Ashby, makes the most of the available space, with a largely open plan ground floor, and the four bedrooms and bathroom laid out to minimise wasted circulation space. Potential problems with head-room in the beamed first floor were cleverly overcome by reducing the level of the entire ground floor by one metre, work undertaken by the owners on a DIY basis, underpinning the structure by a further two metres as they went. A huge amount of DIY effort went into this project and consequently, whilst the cost represents exceptional value for money, it took eight years to complete. This has, however, allowed the owners the luxury of time to pay great attention to detail and as a result, this elegant and spacious home has tremendous character.

A FIRST HAND ACCOUNT OF A BARN CONVERSION

ALL THEIR OWN WORDS

Venita Dickens explains how she and husband Gerry turned a dilapidated timber barn into an elegant contemporary style home.

WORDS: VENITA DICKENS **PHOTOGRAPHY:** PAUL DIXON

IN THE SPRING of 2000 my husband Gerry, who runs his own building and property development company, was asked by a local farmer to accompany him to view a range of dilapidated old farm buildings he was interested in buying for renovation purposes. One of the buildings was a decrepit old timber framed barn, surrounded by a glorious meadow. It had wonderful views over the rolling countryside towards the sea and, although set in a rural position, was only five minutes from the Channel Tunnel.

With full planning permission for conversion to a single dwelling and its convenient location, the setting was near perfect with France clearly visible only 20 miles away. Two years on and following a great deal of hard work and heartache, the barn is now our lovely new home.

The barn was built of softwood pine in about 1880 and used for processing hay and grain. The huge front entrance with overhung canopy had more or less collapsed and there were large gaping holes in the roof. However, the main brick and timber frame structure was in surprisingly good condition.

Gerry is an experienced project manager and chartered Quantity Surveyor and although he had never attempted a barn conversion before, or indeed a self-build scheme, he was well qualified for the job. Although detailed planning consent was already in place, we felt it wise to retain the services of an architect and structural engineer to produce specific working details. Both men were known to us and had specialist knowledge of renovating old buildings.

The barn was divided into five bays, the central one being slightly larger than the other four. On reflection, one of the best pieces of advice we had from our architect was to visit as many barn conversions as possible to ➤

The open riser staircase was made from American white oak, with two curtail steps at the bottom to create a grander appearance.

"ONE OF THE BEST PIECES OF ADVICE WE HAD WAS TO VISIT AS MANY BARN CONVERSIONS AS POSSIBLE TO COLLECT IDEAS."

collect ideas. We were particularly impressed with one where the owner, an interior designer, had decided to maintain the integrity of the external superstructure and not use 'pretend' period features internally. This struck a chord with both of us and we decided to follow suit and dispense with tradition.

The problem generally with barn conversions is that to get the headroom for a first floor, the internal ground level has to be reduced or the roof level raised, both of which involve substantial costs. We decided the most economical solution was to raise the internal roof structure by 600mm, which would give reasonable floor to ceiling levels. In addition we decided to use a separate, independent, internal load bearing steel structure for the first floor to avoid placing extra weight on the existing brick/stone plinth. As a result this negated the necessity to underpin the existing substructure, saving us considerable time and expense. It was necessary, however, to remove the existing internal concrete floor and relay a new reinforced concrete slab and damp proof membrane with internal thickenings to support the steel columns and beams used for propping up the first floor.

In May 2000, construction work began with the erection of a new but sympathetically designed detached garage/log store which could double up as covered site storage. A new water main had to be installed from the nearest service point, approximately 600 metres away. Luckily our friendly farmer owned the land and granted us permission to install the water pipe.

The new internal concrete floor slab provided us with an excellent level platform for internal access to work on the barn superstructure. A temporary hard surface was laid around the outside of the barn for external access and scaffolding. Once we had completed all this work, the old external cladding and roof covering were removed, exposing the skeleton of the original barn.

As the barn is on the periphery of an old Battle of Britain aerodrome, a certain amount of war damage had taken place and sections of the natural slate roof had been replaced with asbestos slates. We had to make special provision – both spatially and financially – for dedicated skips to safely remove the contaminated material to a licensed tip. ➤

One of the major talking points for visitors is the 4ft aquarium overlooking the living area. It required extra timber supports to be incorporated within the landing floor.

The toughened glass panels used for the balustrading needed special stainless steel brackets, which alone cost £400.

It was an awe-inspiring sight to see the barn stripped and laid bare, completely exposed and unprotected. Unfortunately gale force winds arrived and howled straight through the centre of the structure, together with torrential rain. Fortunately, we had the foresight to provide temporary bracing prior to the removal of the external cladding and so the structure survived two violent storms.

Once the skeleton had been exposed, the structure was sandblasted. It took a specialist company two days to complete this operation using a mobile telescopic platform and the weather (for a change) was perfect. In addition the brick and stone plinth was sandblasted inside and out as we had decided to leave this exposed. We also took the opportunity, whilst clear access was available, to have timber repairs carried out, together with the installation of the structural steel work supporting the first floor. The whole of the outside was then scaffolded to enable timber repairs to the external walls to be completed. The main structural alterations could then begin, which meant raising the lateral tie beams by 600mm. A birdcage scaffold was erected internally with a continuous boarded platform below the level of the beams. In addition, a loadbearing scaffold had to be erected above the beams and strengthened, to cope with the weight of hand-winching the timbers, which we estimated weighed approximately one ton each!

The claret colour Rayburn contrasts strikingly with the black slate floor and the units are a Shaker style in limed oak with extra deep worktops in pippy oak.

Luck was not on our side and not only did we have to do battle with the wettest autumn on record, but Gerry had a severe recurring back problem. This created major problems with the programme whilst constant, torrential rain also prevented men from working.

The end of January 2001 arrived and for the first time in approximately six months we were virtually watertight. We could now crack on – one month behind schedule and somewhat independent of the weather – with completing the internal carcassing. We decided to install warm water underfloor heating, which seemed to provide all the answers in terms of comfort and efficiency, given ➤

"WHEN FAMILY AND FRIENDS FIRST VIEWED OUR NEW HOME WHAT THEY SAW LITERALLY TOOK THEIR BREATH AWAY."

that we had a very well insulated building with large open areas.

As we had agreed to go with a fairly contemporary feel to the barn's interior, critical decisions relating to the main focal points had to be decided upon. The carpenter and electrician had a particularly challenging job in solving the problem of how to discreetly install the 4ft aquarium into the galleried landing. Even for such a slim tank it still required 50 gallons of water and extra timber supports to be incorporated within the landing floor to carry the weight. The end result, with the shoal of red and blue neon fish darting about, is like a moving work of art and a real talking point.

Another unusual feature is the end wall in the open plan living area, which has been specially coated in pitted polished plaster containing a pigment of deep hyacinth blue: a fabulous effect not normally applied in domestic properties as it is quite costly, but an effective alternative to an inglenook!

The staircase, being a key feature within the barn, had to be extremely stylish. However, our original curved design proved horrendously expensive to construct and we had to compromise. We chose American white oak with open risers and small square spindles. The two curtail steps at the bottom help create a slightly grander look, more in keeping with a large area. The toughened glass panels used for the ballustrading needed special stainless steel brackets, which alone cost £400.

Landscape architect Mark Baldock was employed to ease the barn into its surroundings.

For a degree of privacy, we divided off the kitchen/ breakfast room and utility from the main living area by using a partition with a portholed swing door. The kitchen, which is a work of art, was custom built by a local joinery company. The claret colour Rayburn contrasts strikingly with the black slate floor and the units are a Shaker style in limed oak with extra deep worktops in pippy oak. The top of the central island unit is made out of black starburst granite and inset with a black halogen hob, covered with a contrasting state of the art stainless steel extractor hood.

When family and friends first viewed our new home what they saw literally took their breath away. When we remember what once stood here, we cannot help but be extremely proud of our achievement. Would we do it again? Ask us again in two years time! ■

BEFORE

FLOORPLAN

To get the headroom for a first floor it was decided to raise the roof by 600mm. This made way for the five bedrooms and three bathrooms. A separate, independent, internal load bearing steel structure for the first floor was installed to avoid placing extra weight on the existing brick/stone plinth. A sympathetically designed detached garage/log store was also included into the design.

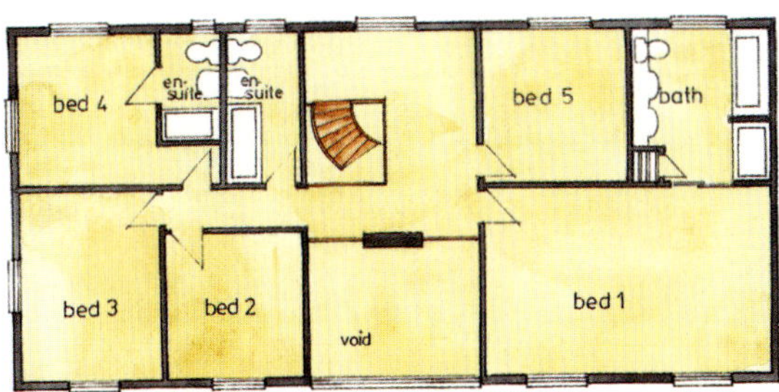

FIRST FLOOR

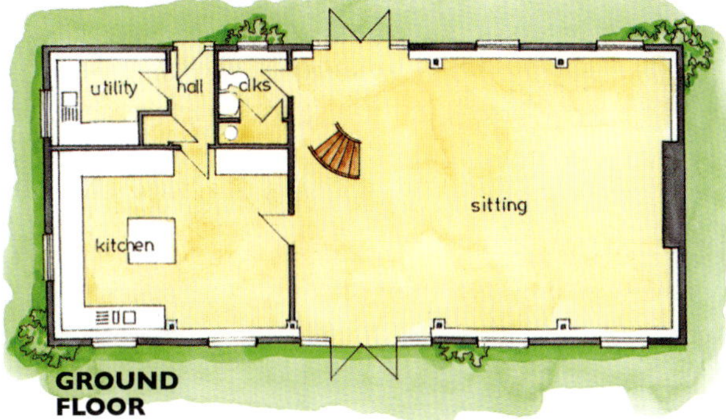

GROUND FLOOR

"TWO YEARS ON AND FOLLOWING A GREAT DEAL OF HARD WORK AND HEARTACHE, THE BARN IS NOW OUR LOVELY NEW HOME."

FACT FILE

Costs as of Nov 2002
Names: Gerry and Venita Dickens
Professions: Chartered surveyor and freelance writer
Area: East Kent
House type: Barn conversion
House size: 305m^2 + outbuildings
Build route: Selves as project managers with subcontract labour
Construction: Pine frame barn and structural steelwork for first floor
Finance: Private
Build time: 12 months
Land cost: £120,000
Build cost: £211,115
Total cost: £331,115
House value: £550,000
Cost/m^2: £692

40%
COST SAVING

USEFUL CONTACTS

Architect – Mervyn Gulvin RIBA: 01227 830881; **Structural Engineer** – KLC Technical Services: 01304 240502; **Timber preservation & DPC Injection** – Homeguard: 01303 255051; **Timber Repair Chemicals –** SBD Ltd:01525 718877; **Structural steelwork/ sundry metalwork** – Dickie Willis:01304 202227; **Roof and Floor slates** – Stoneleaf: 01277 841555; **Specialist Polished Plaster effect** – Armourcoat: 01732 460668; **Limestone Floor Tiles** – The Good Flooring Co: 01892 825675; **Kitchen & Bathroom Joinery** – Thoroughly Wood:01303 863334; **Staircase & Bedroom Joinery** – F. J. Joinery: 01303 223434; **Central Heating Boiler/Cooker** – Glynwed Rayburn:01952 642000; **Underfloor Heating** – NuHeat:01404 549770; **Sanitwaryware** – Ideal Standard:01482 499380; **Windows** – Jeld-Wen UK Ltd:01302 394000; **Stainless steel fixing brackets –** T. Saveker Ltd:0121 3311903 **Alpha Septic Tank** – Klargester Environmental:01296 633000; **Custom made blinds** – Eclectics: 01843 852888; **Landscape Architect** – Mark Baldock: 01304 852415; **Plants –** Blakeney House Nurseries:01227 831800

PRETTY IN PINK

John Walker converted his barn, finished in Suffolk pink, using "as much of the original structure as possible". It's a maxim that will inspire many fellow converters.

WORDS: CLIVE FEWINS PHOTOGRAPHY: PHILLIP BIER

JOHN WALKER DESCRIBES himself as a 'woodaholic'. Fourteen years ago, at the age of 56, he set himself the task of finding an East Anglian barn to convert into a retirement home for himself and his wife, Joan. Three years later, in 1989, he found just what he wanted – a massive 21m long, 370m² oak framed listed Grade II threshing barn, parts of which date from the 16th century, with permission to convert into a house. He purchased it within a few months but it has taken him ten years to complete the conversion.

The barn gradually became transformed into a giant workshop and home to the large array of woodworking equipment he began to acquire. He spent two years planning the conversion of the barn, liaising with planners at Mid Suffolk District Council as he went along.

Pretty soon he realised that he was on the same wavelength as the planners. "My policy from the outset was one of minimal interference with the basic structure of the barn," John explains. "My main concern was to divide it up into living spaces from which one could see and appreciate different parts of the marvellous structure. Once the planners realised that this was my approach – and that I proposed to use oak throughout and not install any large modern partition walls – they seemed pretty happy with what I intended to do."

With the exception of the main guest rooms, bathrooms and WCs, John tried to keep to half height screens to divide the rooms from the rest of the barn. "It is really a version of the open plan style, in order to

The design, which John came up with himself, uses half height screens instead of partition walls to keep the beautiful roof structure to the fore.

"IT WAS THE WONDERFUL ROOF STRUCTURE, WITH ITS CATHEDRAL-LIKE QUALITY, THAT REALLY SOLD THE BUILDING TO JOHN."

The windows are positioned to allow the sun to get into most parts of the barn during the course of the day.

make as much of the magnificent building visible from as many parts of the interior as possible," he says.

Despite his meticulous planning, when local renovation specialist Graham Black and his team got to work, it took them three years, on and off, to turn the building into a habitable space. One of the main jobs was jacking up the entire ground floor structure one bay at a time, as the oak cill plates needed replacing throughout. As a result of this, the brick cill needed a couple of extra courses, as most of the ground floor studs had to be taken out, cleaned up, shortened and remorticed. Some underpinning was also needed.

Another major task was thatching. Long straw specialists Steve and Peter Letch took three months and used seventeen tonnes of straw to complete the task.

John was happy to let Graham be the main contractor because of his local connections and experience with old timber-framed structures. Besides, at the time, John was still busy day-to-day running his Essex-based electrical component distribution business.

Sadly, Joan died in February 1995 after a short period of illness. In July of the same year John sold his business and retired. He had already sold his house in Essex – which provided some working capital for the reconstructed 'east wing' on the site of the former cartshed – and was living in a rented property nearby.

At the beginning of 1996, John was able to move into what he calls "the cupboard" in the barn – the roof void above the new single storey east wing. He stayed there, eating, sleeping and cooking in the same room, until Graham and his team moved out at the end of 1996. Since then he has been fitting out the building, making furniture – including the magnificent oak dining table – and working on the shelving in the library, assisted by new wife Geraldine.

It was the wonderful roof structure, with its cathedral like quality, that really sold the building to John. It is the dominant feature, soaring an extra ➤

To the west of the building, a staircase leads up another flight to John's office, which occupies a third floor gallery.

5m above the 5m cill plates. As John says, it is a breathtaking sight for anyone who loves barns.

The second factor to be singled out is John's design. Although he employed an architectural technician to draw up the detailed plans for submission, he spent many hours at the drawing board coming up with the "galleried" concept, which makes sure that the vast majority of the grand interior of the building can be seen from most of the rooms.

You certainly get the feeling that you are in a complete barn, not something that has been divided up and segmented in the unfortunate way some lovely barns were in the 1960s, before local authorities tightened up and many historic barns became listed buildings.

The design involves twin oak staircases, with triangular solid treads on oak bearers rising from either side of the massive oak front door and enclosed hallway. A narrow landing joins the two sides of the building, passing across the gable ends of the former threshing bay. The upper half of this houses the library, which will eventually double as a guest room.

"Constructing the staircases so there was room to walk at first floor level, beneath the huge curved oak braces at the side of the building, was very tricky," recalls John. "As it was such a squeeze, we just had to make it fit as we went along. Graham was admirable to deal with and talked it through with me all the time.

Even the master bedroom set on a first floor gallery is open to the main living area below. At night, the space can be curtained off for privacy.

"Graham was also meticulous about the positioning of any new vertical posts erected to support the new first floor. To avoid disturbing the rhythm of the building, he carefully positioned these so they lined up with the original main studs on the east and west walls."

To the west of the building, a staircase leads up another flight to John's office, which occupies a third floor gallery. Here he keeps his drawing board and pursues his favourite hobbies, of calligraphy and bookbinding, and Geraldine paints. The room has a bird's eye view over the main space of the barn, the 'wall' on that side being a half height oak screen. This room also doubles as a spare bedroom, when needed.

The master bedroom beneath is constructed in the same manner, with a half height oak screen rather than walls separating it from the main soaring central open space of the building. At night, when John and Geraldine are asleep, it can be curtained off.

Across the landing at the far end of the building is another third floor gallery, accessed by a ladder and only really used when John and Geraldine hold charity concerts in the barn at Christmas. It is a good vantage point ➤

The single storey east wing – formerly a cartshed – was the first section to be completed. John moved into the roofspace and lived there for three years during construction.

to climb to and survey the full extent of the building. It is also an ideal place from which to view the construction of the windows, of which John is justifiably proud.

The entire upper end is glazed. Looking up at the windows, it is possible to work out how John managed to design so much light into the building without disturbing any of the original fabric or making the conversion look unbalanced.

The lower, unglazed sections of the gable ends are clad with chestnut weatherboarding treated with linseed oil. "Every oak window unit was purpose made by JSR joinery of Great Yeldham, Essex, to fit the opening it was designed for," says John. "The key was to select the openings so that they created a completely irregular effect, while importing sufficient light. We had to ensure that the units could be fitted on the outside of the main frame of the barn so that only the original framing can be seen from the inside.

"One of the great joys of the barn is the way the windows are positioned. The sun lights most parts of it during the course of the day. The different sections of the oak take on an even more golden glow than usual," he enthuses. "It adds greatly to the experience of living here. People say the barn has a relaxing atmosphere. I think that the sun shining in, illuminating odd sections in a random pattern, is partly responsible for this."

The ground floor is one of the other great successes of this striking conversion – the floor of Norfolk pamments was made from clay dug from the field next to the barn. Each of the nine inch square pamments, hand made by Norwich craftsman Julian Burgess, has a natural, slightly varying hue, from pale pink and pale creamy yellow to subtle terracotta. There are even hints of blues, found naturally in the clay. Again, they come into their own when the sun catches a patch of them as it shines through one of the many windows.

"WHERE ONE OR TWO OF THE ORIGINAL WATTLE AND DAUB PANELS SURVIVED, THEY HAVE BEEN PRESERVED AND INCORPORATED INTO THE NEW PLASTER."

The floor is a good example of what John calls 'looking at what you have got and keeping the theme'. "A good maxim for converting any old building is 'try, all the time, to work with the structure rather than against it'," he says. "Another phrase Geraldine and I use a lot when we are looking at an addition or alteration to the structure is, 'is it barn friendly?'"

Drawing on what was in the locality, John has opted for a 'Suffolk pink' for the limewash on the outside of the building. It covers a lime plaster attached to a stainless steel mesh, in turn attached to plywood sheets that hold in 100mm sheets of foil backed polyurethane insulation, positioned between the oak studs. Inside, the insulation is covered by more stainless steel mesh, attached to the inside of the studs by staples. The internal panels are plastered with a Carlite Bonding and lime mix, which has also been used on the ceilings. Where one or two of the original wattle and daub panels survived behind the black steel sheeting, they have been preserved and incorporated into the new plaster.

"Most of the original studs are about 8 inches deep, affording ample space to leave half of them exposed on the inside, so all the original structure can be seen," Graham points out. "It is an effective technique. As well as providing a high standard of insulation, it has given the plaster the elasticity to move with the building."

Above all, it is the beauty of the oak that leaves the most lasting effect at The Thatched Barn. In all it is an astonishing and inspiring conversion, a genuine tour de force. However, John insists that it is the barn and not him that deserves most of the glory. "It is a noble building and I like to think it has gained something at a crucial time in its history as a result of my efforts," he says. "Although I technically own the barn, Geraldine and I are in reality just caretakers. It is the building that really matters." ■

FACT FILE

Costs as of Aug 2000
Names: John and Geraldine Walker
Professions: Retired
Area: Suffolk
House type: Oak framed barn conversion
House size: 370m²
Build route: Self as main contractor
Construction: Oak frame with infill panels
Finance: Private
Build time: Ten years
Barn cost: £100,000
Build cost: £250,000

House value: £400,000
Cost/m²: £675

12%
COST SAVING

FLOORPLAN

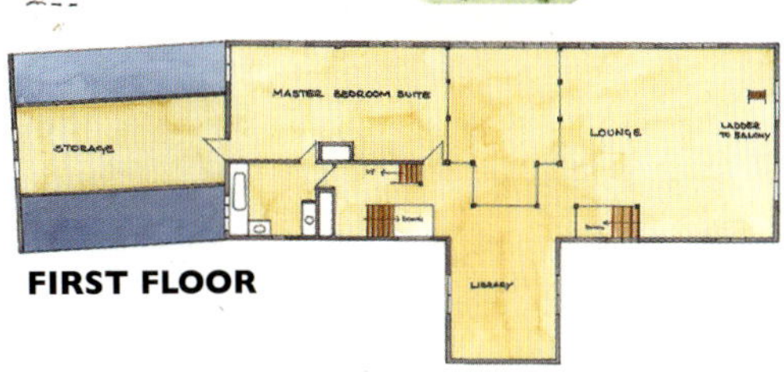

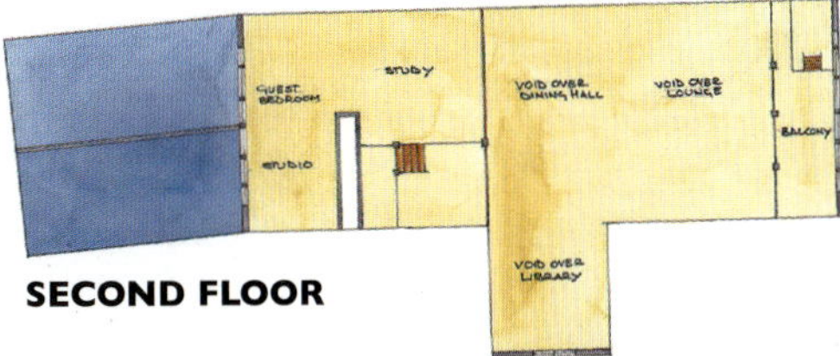

The first floor contains the lounge and master bedroom suite whilst on the second floor there is a guest bedroom/study/ studio and eaves storage area over the garage

USEFUL CONTACTS

Window Frames – JSR Joinery: 01787 237722 **Carpenters** – Ralph Oxburgh: 01379 741544 Tony Holyoak: 01379 384597 **Blacksmith** – John Jarvis: 01379 644319 **Ironmongery** – Ashfield Traditional: 01449 711273 **Thatching** – Steve and Peter Letch: 01379 852335 **Timber sawing** – On Site Services: 01284 386056 **Underfloor heating** – Wilmington Products: 01392 444122

CONVERTING OUTBUILDINGS INTO LIVING ACCOMMODATION

LINKING UP

Aaron and Suzanne Evans have converted outbuildings into an elegant living space, thanks to a clever architectural solution.

WORDS: DEBBIE JEFFERY
PHOTOGRAPHY: JEREMY PHILLIPS

The cottage has been shoehorned onto a narrow strip of land which had previously been used for storage. The stable outbuilding has been re-roofed with the original clay tiles – pantiles on one side and double Romans on the other.

SHEEPFAIR BARN WAS certainly a family affair," explains architect Aaron Evans. "We bought an old barn which we converted in 1998, and some years passed before we decided to tackle the group of outbuildings which stood opposite. Our younger daughter Sarah was studying to be an architect, and we asked if she would like to design a small conversion as an opportunity to get some hands-on experience, with her mother, Suzanne, taking on the role of client."

Originally, planning permission and listed building consent for the barn included approvals for converting the small stone stable and removing the garage, but it soon became apparent that the garage was built into the neighbour's wall, and that demolishing it would be physically impossible.

"We wanted to use the outbuilding as guest accommodation," says Suzanne, an illustrator, "but due to the site constraints, a new solution was required. Sarah and Aaron decided to double the size of the stable by linking it to the garage with a flat-roofed addition, creating a two bedroom cottage and private courtyard, which has lovely southern views across open fields towards Bath."

Negotiations with the local authorities were protracted due to the sensitive nature of the site. Marshfield is the southernmost village of the Cotswolds, with a lively mixture of cottages and Georgian buildings set in a Conservation Area.

Sarah's proposal to expand the scheme by linking the garage and outbuilding with a flat-roofed construction was finally accepted by the

"THE CHALLENGE WAS TO MAKE AN INTERESTING STAIRCASE IN A SMALL SPACE."

Feature staircase
The oak-clad steel and concrete staircase appears to float without support, and is lit from above by a skylight and from below with floor-mounted spotlights.

planners, who had initially suggested that the new link should match the existing vernacular architecture. Sarah successfully argued that the flat roof would be almost invisible between the two pitched roofs, and also worked on the building regulations. These were challenging due to the fact that the outbuildings border two other boundaries, and adjoin the wall of a property which had to be damp proofed prior to building a new cavity wall for the cottage.

After working on the project for almost a year, Sarah had to go back to college and Aaron oversaw the building work on site, employing the same building contractor he had used for the barn conversion. "Emerys of Bath are knowledgeable local builders with a family tradition going back over many years. When I came to Bath my first project involved working with the grandfather," says Aaron. "Since then I've worked with the son and now the grandson, Simon."

The builder removed the collapsed roofs and took down some of the old stone and blockwork walls, before digging out the ground levels. These were crucial to avoid the new dwelling overlooking or undermining the neighbouring properties, and were checked on a daily basis. Much of the Hamstone used for the external leaf of the cavity walling was recovered from the site, with additional reclaimed stone sourced by the contractor who laid it in courses using a lime mortar. The original stone outbuilding had a curved wall, which has been reflected internally in the sitting room and the bedroom above.

Timber joists bear on a new party wall, formed from the garden wall of the neighbouring property, which had to be strengthened. These joists span across to an exposed galvanised RSJ making a strong feature of the entrance elevation, which is shielded from the afternoon sun by a simple oak brise-soleil. The flat roof has been covered in lead, while the ➤

Grey units from Ikea combine with fitted maple worktops and simple glass splashbacks in the compact kitchen.

"IT IS A FACT OF VILLAGE LIFE THAT EACH BOUNDARY IMPINGES ON ANOTHER, WHICH ALSO MAKES CONSTRUCTION INTERESTING AND CHALLENGING."

two older structures were repaired and re-roofed using the original clay tiles from the stables with reclaimed slate tiles on the garage.

The interiors have been finished as simply as possible, without the addition of skirting boards, architrave, picture rails or coving, which Aaron felt would be inappropriate for a rural agricultural building. Instead, he asked the builder to plaster the internal walls to create a recessed joint, or shadow gap, at the base of the walls and around the doors and windows for definition and a more contemporary finish.

When it came to designing the staircase in the sitting room, however, Aaron and Sarah were determined to express themselves and have some fun. "We wanted a real statement," says Sarah, who has now graduated and is working as an architect in London. "The challenge was to make an interesting staircase in a small space, and we came up with the idea of 'floating' it from the top to the bottom without touching the sides – allowing light to pass down and create a shadow on the wall. Two sheets of glazing form a balustrade and floor-mounted spotlights create a glow at night." The steel stairs and concrete winder were clad in oak by a joiner, who also made the windows and internal doors. Although the £135,000 budget imposed certain restrictions, the family managed to include luxury finishes such as the travertine stone floors over underfloor heating and the gravity-defying oak staircase.

"Over time, we hope that the cottage will mellow and become like other buildings in the village," says Aaron. "It is a fact of village life that each boundary impinges on another, which also makes construction interesting and challenging. The cottage has been shoehorned into a very tight, narrow site, but stumbling upon little tucked-away buildings is all part of the charm of these places." ■

FACT FILE

Names: Aaron and Suzanne Evans
Professions: Architect and illustrator
Area: Wiltshire
House type: Two bed cottage
House size: 108m^2
Build route: Main contractor and subcontractors
Construction: Stone and block walls, slate, clay tile and flat roofs
Finance: Private
Build time: Oct '03 – May '04
Land cost: already owned, est. value: £85,000
Build cost: £135,000
Total cost: £220,000
House value: £250,000
Cost/m^2: £1,250
Cost Breakdown:

74%
COST SAVING

Demolition	£5,200
Floor slab/foundations	£5,570
Rubble walls	£6,012
Lining wall incl. render	£8,492
Copings	£2,000
Partition walls	£2,564
Roof structure	£5,920
Roof coverings	£6,941
Rainwater goods/brise-soleil	£3,273
Ceilings	£2,532
First floor	£2,354
Staircase	£5,500
Glazed link	£5,062
Windows/external doors	£2,522
Internal doors	£6,600
Sanitaryware	£2,700
Kitchen	£4,200
Mechanical	£9,240
Electrical	£8,750
Incoming services	£1,550
Decoration	£4,830
Stone floor (supplied/laid)	£4,500
Tiling	£1,550
External paving	£1,830
Landscape/garden walls	£9,800
Drainage	£2,440
Prelims	£12,719
TOTAL	**£134,651**

FLOORPLAN

A two storey stone outbuilding contains the sitting room with a bedroom and bathroom above, and the former garage is now a single storey studio bedroom with an en suite bathroom. Between these two buildings a new flat-roofed link houses the kitchen/diner.

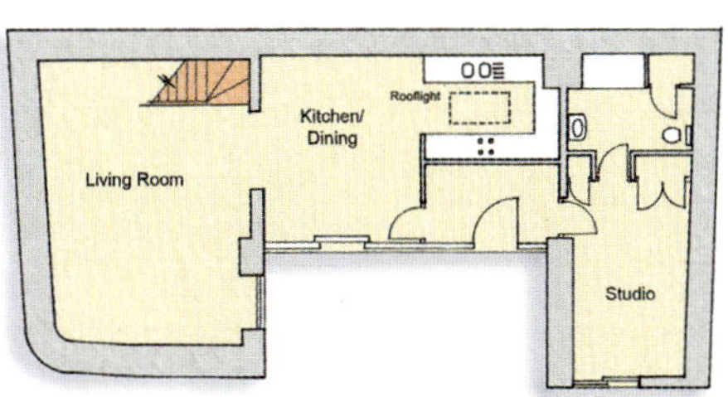

GROUND FLOOR

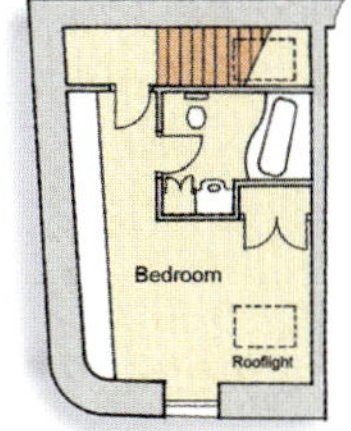

FIRST FLOOR

USEFUL CONTACTS

Architect – Aaron Evans Architects Ltd: 01225 466234; **Building contractor** – Emerys of Bath: 01225 462153; **Electrical contractor** – Darren Wilkins: 01761 415310; **Plumbing and heating** – TPM South West Ltd: 01225 851167; **Kitchen** – Ikea: 020 8208 5600; **Underfloor heating** – Wirsbo: 01455 550355; **Flat roof membrane** – Sarnafil Ltd: 01603 748985; **Sanitaryware** – Ideal Standard: 01482 346461; **Stairs and windows** – BDF Joinery: 01749 345395; **Stone floor tiles** – Mandarin: 01225 460033

CONTEMPORARY STAIRCASES

Designers of contemporary style staircases tend to favour the less-is-more approach, creating a minimal structure that is elegantly simple in its form. The staircase in this property follows this ethos, taking the form of a traditional staircase comprising of oak treads and risers, but without the conventional strings that run up either side to hold the staircase together, or the conventional balustrading arrangement of newels, spindles and handrail. Positioned 100mm away from the adjacent flanking wall and with only two sheets of toughened glass balustrading to the other, the staircase has no visible means of support other than at the top and bottom. Cleverly concealed within the oak treads and risers are two structural 'strings' made from sections of mild steel. For more on staircase design, visit www.homebuilding.co.uk and search for 'staircases'.

CONVERTING AN OLD OAST AND ADJOINING BARNS

The building comprises three roundels attached to a two storey barn plus a second barn annexe, attached via a contemporary style glazed walkway.

John and Sarah Garthwaite have turned an unusual disused oast house and adjoining barn into a unique contemporary style family home.

WORDS: DEBBIE JEFFERY
PHOTOGRAPHY: ANDREW SOUTHALL/ ARCBLUE.COM

ANOTHER ROUND

JOHN GARTHWAITE AND Nicholas Kidwell first met at school and, while Nicholas planned to become an architect, John hoped to be able to employ his friend to convert the oast in the grounds of his parents' home. Both boys got their wish, and the result is a sensitive conversion which links the existing oast to a listed barn.

"We moved from London in 1988 and bought an Elizabethan house in a village in Kent," Sarah Garthwaite explains. "John and I renovated the house, which was great fun and gave us a real insight into this kind of project. When Matfield Oast became available we were ready for the challenge – and knew Nick would be our architect."

Oasts were originally constructed as large kilns to dry hops for making beer, which were loaded at the top of the funnel with a fire set beneath. Matfield Oast is a brick and timber framed building with a weatherboarded first floor and traditional Kent peg tiled roof. Last used as a working oast in the early part of the 20th century, the building had been owned by the Garthwaite family since 1960.

"John and I had always thought that it would make a wonderful home and approached the other members of the family, who agreed to sell us the property and two acres of land," says Sarah. "It was completely unconverted, used for storing tractors, and the hop presses had been removed – but everything else was intact."

A semi derelict barn has been repositioned and connected to the main oast with an elegant glazed steel and oak structure, with frameless double glazing and a copper roof.

Most oast houses are situated in a group of other, often relatively unattractive buildings, but Matfield Oast comprises three roundels attached to a two storey barn and stands alone in the middle of an apple orchard. "It had good sized internal spaces and a lot of potential," says John, who had – quite coincidentally – previously worked as a hop farmer.

He and Sarah had also acquired a small 17th century barn, which stood in the orchard. The decision was taken to physically move this semi-derelict structure 80 metres and link it to the main group of buildings with a glazed pergola, extended to enclose a formal herb garden.

"The idea was to avoid breaking up the internal space of the main two storey barn with too many subdivisions," says Nick, who proposed to ➤

Oak was used for the treads and handrail of the staircase, which features perforated stainless steel balustrading.

"I INVITED THE CONSERVATION OFFICER TO SKETCH SOMETHING OUT THAT HE WOULD APPROVE, WHICH FLUMMOXED HIM!"

organise the house around the full height central roundel. "By incorporating the single storey barn as a playroom for Sarah and John's two children it became possible to reduce the number of rooms in the main oast –allowing spacious living spaces. Diagonal walls in the dining hall both open up this area to the staircase and create smaller corner 'service' rooms such as the utility, bathrooms and airing cupboard."

With much local scepticism about the proposed project Sarah, a GP, decided to lobby every single councillor on the planning subcommittee, writing a detailed letter explaining their aims, quoting chunks of planning policy guidelines and inviting them to visit the site. "I'm extremely interested in architecture and find it fairly easy to visualise what can be done with spaces without the need for plans or 3D drawings," she explains. "At the site meeting I invited the conservation officer to sketch something out that he would approve, which flummoxed him!"

The central roundel, with its gravity-defying cantilevered staircase, houses the dining hall – featuring an elm table fitted with fibre optic lights.

Gaining planning permission for the design took almost two years, with the conservation officer determined that, if the grade II listed building was going to be extended, it was important to make a clear distinction between the old and the new. "His preference for a contemporary structure gave us the opportunity to design something really unique," says Nicholas Kidwell of the curved link – which has frameless double glazing, powder-coated steel doors, a copper roof and oak supporting columns.

Sarah designed the kitchen units herself and sourced cut price granite worktops from the local memorial mason.

The Garthwaites remained in their previous home during the build – managing to co-ordinate the sale with moving into the oast – and Nicholas Kidwell was employed to manage the entire project. Following the tender process, the contract was awarded to Hertfordshire contractor MJ Wadey, who had worked with Clarke Kidwell Architects on previous occasions. "It was important to us not to involve too many subcontractors," says John. "We wanted a company who could handle most of the work themselves, so that they were absolutely in control."

Extensive site preparation included demolishing the existing partitions within the building, which had once housed animals. Structurally, the oast proved to be quite sound, although minor subsidence was detected in one ➤

The informal living room, situated in the original barn building, features a contemporary style fireplace.

corner and, as part of the renovation, it was necessary to underpin the entire building as it had no real foundations.

"We also later discovered that some of the main purlins were unsupported – which meant that the roof needed to be stripped right back and rebuilt along about one third of its length," recalls Sarah. "This turned out to be a blessing in disguise, as it then became cost-effective to incorporate a large rooflight into this section of roof, which totally transforms the interior. I can't imagine what the house would be like without it."

With the roundels built in nine inch solid brickwork it was necessary to add insulation and a new inner blockwork skin at ground floor level to meet current building regulation requirements and to support the cantilevered stairs. The conical upper parts of the roundels have been dry lined and painted white.

It took a great deal of preparation before the small barn was ready to be moved. Tiles were stripped from the roof and the entire building braced with steel girders to ensure that it remained intact when lifted from its plinth. A new base plate and brick plinth were constructed, which raised the barn slightly in order to cleanly meet the roof of the curved link building. "We all held our breath while the crane manoeuvred the barn across the orchard," says Sarah. "It could so easily have gone wrong!"

Once in position, the barn was refurbished to expose the original internal frame and elm boards. External oak boarding was salvaged from storm-damaged trees, and reclaimed Kent peg tiles purchased from a variety of sources to re-roof the structure. Now an open plan playroom, the single storey barn leads into the glazed link and allows the living room to retain the original large barn door opening, fitted with new folding doors.

The most imposing space in the conversion is the central roundel, which houses a cantilevered steel and oak staircase leading up from the double height dining room. The glazed ridge allow views of the cowl above, which still adjusts in the wind. ■

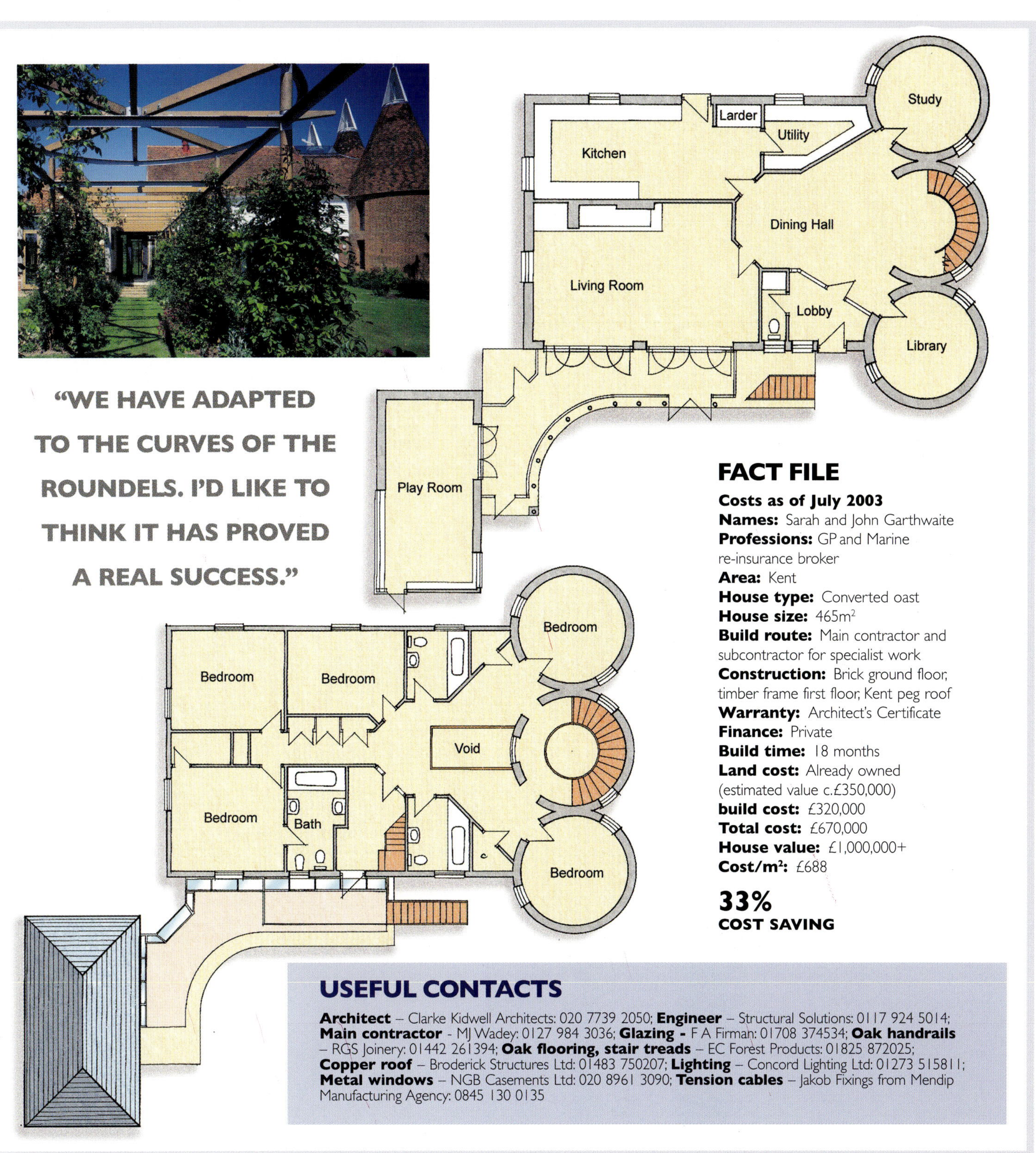

"WE HAVE ADAPTED TO THE CURVES OF THE ROUNDELS. I'D LIKE TO THINK IT HAS PROVED A REAL SUCCESS."

FACT FILE

Costs as of July 2003
Names: Sarah and John Garthwaite
Professions: GP and Marine re-insurance broker
Area: Kent
House type: Converted oast
House size: 465m²
Build route: Main contractor and subcontractor for specialist work
Construction: Brick ground floor, timber frame first floor, Kent peg roof
Warranty: Architect's Certificate
Finance: Private
Build time: 18 months
Land cost: Already owned (estimated value c.£350,000)
build cost: £320,000
Total cost: £670,000
House value: £1,000,000+
Cost/m²: £688

33%
COST SAVING

USEFUL CONTACTS

Architect – Clarke Kidwell Architects: 020 7739 2050; **Engineer** – Structural Solutions: 0117 924 5014; **Main contractor** - MJ Wadey: 0127 984 3036; **Glazing -** F A Firman: 01708 374534; **Oak handrails** – RGS Joinery: 01442 261394; **Oak flooring, stair treads** – EC Forest Products: 01825 872025; **Copper roof** – Broderick Structures Ltd: 01483 750207; **Lighting** – Concord Lighting Ltd: 01273 515811; **Metal windows** – NGB Casements Ltd: 020 8961 3090; **Tension cables** – Jakob Fixings from Mendip Manufacturing Agency: 0845 130 0135

The main opportunity for letting in light was through what would originally have been the two pairs of barn gates. One end of the barn required rebuilding, however, so the Eastwoods used the opportunity to introduce additional openings.

CONVERTING A BRICK BUILT BARN

How a successful barn conversion changed the lives of Hampshire couple John and Josie Eastwood.

LET THERE BE LIGHT

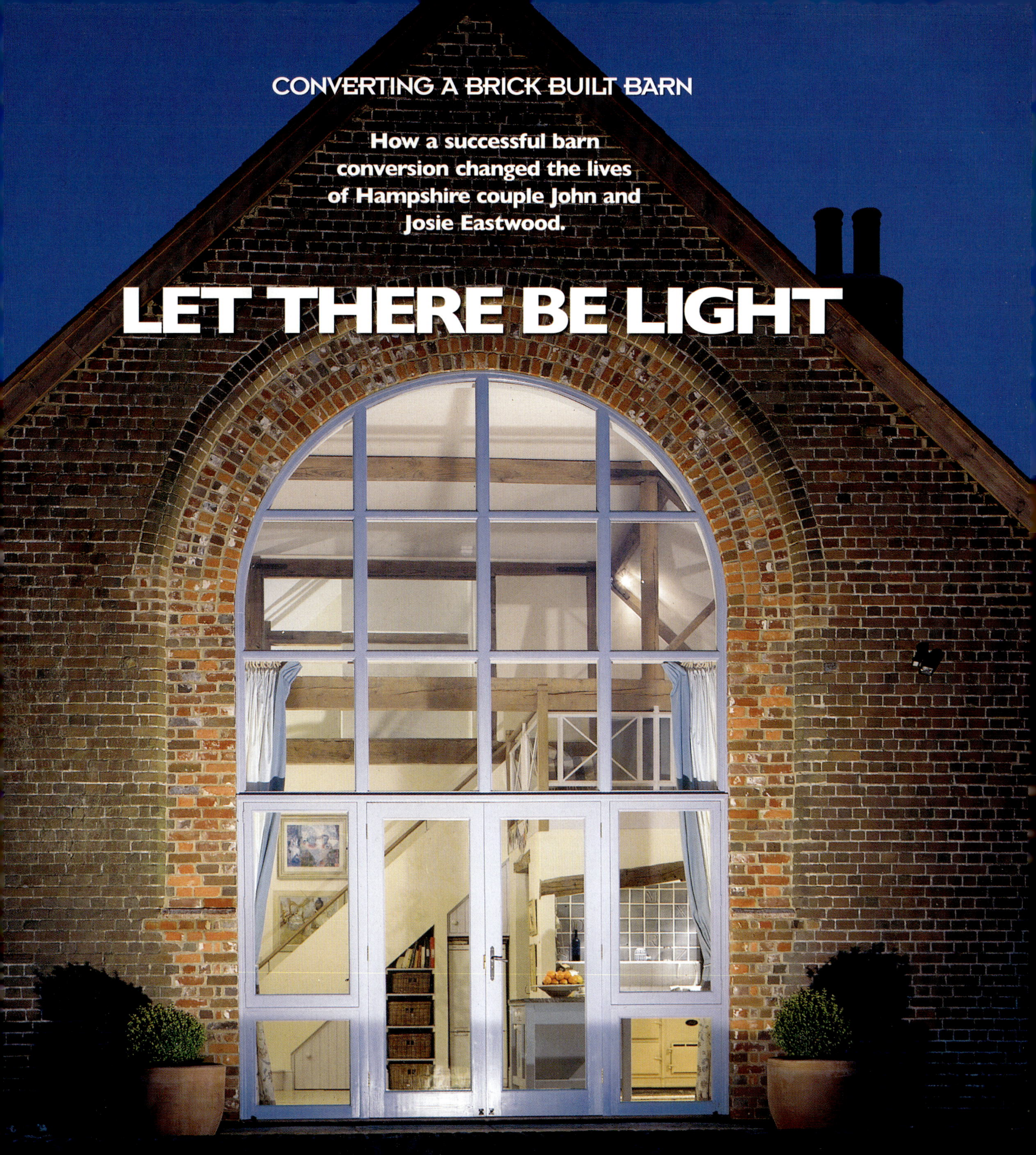

WORDS: JUDE WEBLEY. PHOTOGRAPHY: PHILLIP BIER

WHEN I FIRST saw our farmhouse in Hampshire 20 years ago it was the complex of brick and flint outbuildings which attracted my attention," says Doug Palmer. "I felt sure that, one day, these would be easy to convert into a fantastic home. Little did I realise that this would turn into a DIY job of gigantic proportions. I didn't dare say anything to Suzy – I saved that for the day after we moved in!"

The stresses and strains of running a successful restaurant business, together with bringing up a young family, meant that the couple waited for some years before commencing the conversion project. During this time they renovated and extended the farmhouse, landscaping much of the surrounding five acres. The site had previously been used for car repairs and as a small petrol station and transport café, leaving the whole area littered with rusting remains of vehicles, old petrol pumps and piles of scrap.

Although the Palmers had given much thought to the eventual layout and design for the barn, finding an architect to interpret their ideas and come up with a suitable scheme proved taxing.

"There were several factors which had to be reconciled," says Doug. "Suzy was insistent that the kitchen should have the same southerly view down the valley that we already enjoyed from the farmhouse and, as the barn faces east-west, this seemed somewhat difficult. Secondly, much of the main barn was not high enough to accommodate two floors and, lastly, it appeared that in order to raise the insulation levels to the required standard we would have to dry-line the existing walls – which would

The Eastwoods regret not installing underfloor heating – they feel constrained by radiators on the walls when positioning furniture.

"YOU NEED TO DECIDE HOW MUCH DIFFERENCE A PARTICULAR FEATURE OR PRODUCT IS GOING TO ADD TO THE LOOK AGAINST THE EXTRA COST IT WILL INCUR."

➤

"WHEN BUYING RECLAIMED BRICKS AND ROOF TILES, IT'S VITAL TO CHECK RIGHT INSIDE THE PILE TO ENSURE THAT THE HIDDEN ONES ARE NOT BROKEN OR OF POOR QUALITY."

have taken away the very solid and rustic feel we so wanted to keep."

Ultimately, the problem of the kitchen was solved by adding a single storey pitched roof extension at right angles to the main barn, which not only provides wonderful southerly views but also encloses a sheltered terrace area and gives added width to the interior of the property. The dilemma of retaining a solid 'original' feel to the outer walls was resolved by building an inner skin of insulated reclaimed brick and block.

Structural engineer and family friend, Ian Calder, came up with a solution to the lack of headroom in the main barn. "Ian told us that we could get the extra height we needed by digging out about one metre of the existing floor. The bad news was that the walls didn't go down below ground level, and we would have to sit the entire barn on a ring of concrete extending down nearly two metres into the ground!" explains Doug, who by this time had sold his restaurant business, providing funding and giving him the time to work on the build.

A local bricklayer was employed on a labour-only basis to construct the new kitchen extension and build the chimney in the sitting room. Electricians and plumbers were hired when they were needed, and retired train driver, Cyril Gardner – who had already built his own house in a nearby village – was recruited to help with all the general building tasks. Together, Doug and Cyril tackled every aspect of the build including foundations, drainage, brick and flintwork, carpentry, roofing, leadwork, plastering and glazing.

"There were times when we disagreed about how to do something and it would end up like a scene from Laurel and Hardy," laughs Doug. "On the other hand, Cyril's help was invaluable and it would have been very ➤

The beautiful distressed limestone floor is from Classical Flagstones of Bath.

The double height farmhouse kitchen is overlooked by a mezzanine floor.

"THERE WERE TIMES WHEN WE DISAGREED ABOUT HOW TO DO SOMETHING AND IT WOULD END UP LIKE A SCENE FROM LAUREL AND HARDY."

difficult to work entirely alone. Having someone arrive on site each day at 8am made me stick to the task and put in full days."

Once the underpinning was complete, the internal floor was excavated, drainage laid and a new concrete floor poured. An insulated internal skin, constructed from reclaimed bricks and insulation blocks, was then built inside the original perimeter wall. Most of these original walls are constructed from chalk blocks with brick quoins and reveals, and Doug has replicated this in the new inner shell. Steel joists were then heaved into position – spanning the full width and running the entire length of the building. These pass over the inner skin wall and are supported by the outer walls, and Doug has inserted reclaimed pine beams under each one.

The family eventually moved into the barn at the end of 1999, five years after the start of building. Landscaping followed and was a massive undertaking. "As the ground floor was now nearly a metre below the outside ground level this had to be addressed all around the property," says Doug. "Fortunately I had invested in a compact tractor, with a front loader and mini digger attachment, and with this I was able to excavate, level and re-grade the land before building retaining walls and steps where necessary."

Doug then turned his attention to converting an attached stable building into a guest suite, which necessitated rebuilding two of the walls and replacing the entire roof. This area also provides a garden room, which links the main house to the guest suite. "We particularly wanted to keep this part of the building in its original state," says Suzy. "It's the only section that still had its Victorian brick floor so, by making it into a garden room, we avoided the need to change it in any way."

"I would definitely tackle any future building programme in a different way," says Doug. "I don't think I could face another eight year project! When I started this I was under the impression that 'self-build' meant that you had to build it yourself, but now I've heard of something called 'delegation' and I'm sure that's the way to go!" ■

FACT FILE

Costs as of Dec 2000
Names: John and Josie Eastwood
Professions: Project Manager, Mother/Artist
Area: Winchester
House type: Brick barn conversion
House size: 418m²
Build route: Self project managed + sub contractors
Construction: Traditional
Build time: Seven months
Property cost: £65,000
Land cost: £170,000
Cost/m²: £774

FLOORPLAN

The tendency of barn conversions to end up with one long narrow corridor upstairs from which you reach a series of bedrooms has been avoided by having two staircases to improve access to the first floor.

FIRST FLOOR

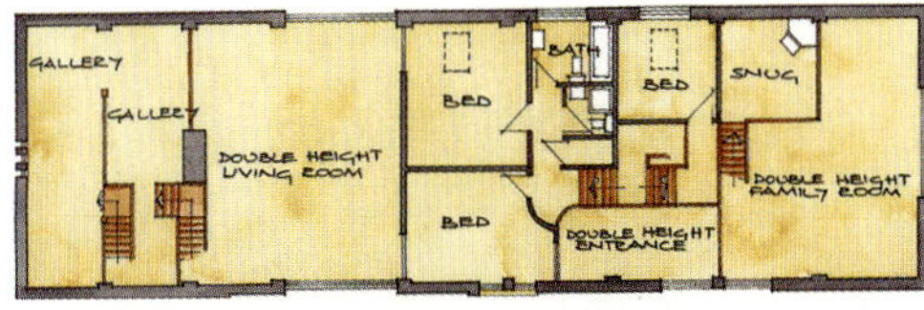

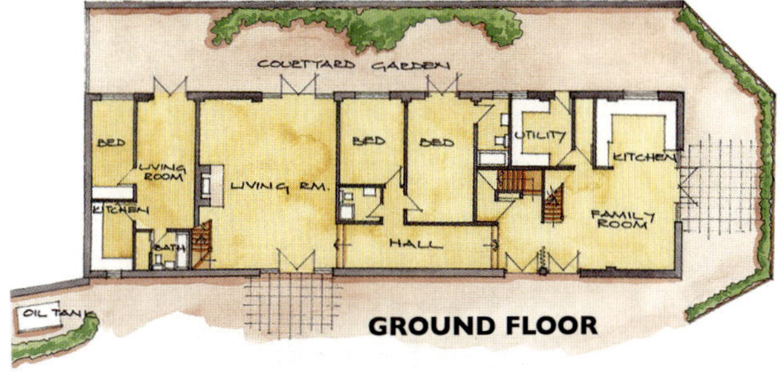

USEFUL CONTACTS

Architect – David Ashe, CH Design (Winchester): 01962 841404 **Builder** – CBco: 0799 056 5711 **Reclaimed materials** – Dorset Reclamation: 01929 472200 Minter Reclamation: 01202 828873 Drummonds Reclamation: 01428 609444 **Ironmongery** – Clayton Munroe: 01803 762626 **Joinery** – SB Joinery: 01264 334106 **Fireplace** – Andrew Calvert: 01969 622515 **External flagstones** – Ruscrete: 023 8086 5046 Tiles – Froyle Tiles: 01420 23693 **Stair Rods** – Roger Oates: 01531 631611 **Paint** – Farrow & Ball (various stockists) **Wallpaper** – Cath Kidston: 020 7221 4000 **Kitchen Floor** – Classical Flagstones: 01225 316759

CONVERTING A BARN ON A BUDGET

SALVAGE STYLE

An eclectic approach to gaining materials helped a Devon couple to convert a barn into a stylish home on a very modest budget

WORDS: DEBBIE JEFFERY PHOTOGRAPHY: NIGEL RIGDEN

BUDGET' IS DEFINITELY not a word which springs to mind when you enter Langford House. This converted barn oozes understated taste and effortless style, but cost only a fraction of its current value to build. Imagination and a great deal of hard work have transformed what started out as junk into the features which define Richard Prowse and Melanie Leach's new home.

Council kerb stones, costing just £40 for three, make up the sturdy stone fire surround in the sitting room; oak timbers have been salvaged from a local tip and sandblasted to be used above window openings and the stylish ball and claw foot bath was once a muddy cattle trough. His work as an arborist, as well as piloting hot air balloons, takes him all over the South West, enabling him to source materials in the most unlikely of places.

The barn is situated on his family's farm and Richard previously converted the adjacent octagonal outbuilding into a small and quirky bachelor pad when he was at art college in the 1970s. "Such hands-on building experience meant that, when I met Melanie, the decision was taken that we needed a larger home and should convert the long barn," he says.

Designed to face east across a meadow and arboretum – part of 17 acres of land the couple now own – the conversion was planned with the help of Richard's friend, architect Peter Thompson.

The century old stone barn was originally on two storeys but, by raising the roof slightly, an additional level was made possible which dictated the entire layout. Two bedrooms, a

The farmhouse style kitchen is not overtly rustic – supplied by Jim Bond, it has an open, modern feel.

"TO KEEP COSTS DOWN RICHARD PURCHASED ALL MATERIALS AND LABOURED ON ALMOST ALL ASPECTS OF THE BUILD."

➤

A chunky staircase, constructed from macrocarpa, leads up from the dining room which, with its glazed roof, has the feel of a conservatory.

bathroom and utility room at lower ground floor level create a quiet self-contained unit for guests, which could be converted into a flat at a later date. The ground floor, accessed from the upper garden, houses the kitchen, sitting room and a galleried dining room roofed in glass – with two further bedrooms, a bathroom and dressing room situated on the first floor. "The glass roof gives us warmth and light even on the wettest days, and is fantastic on clear nights," says Melanie.

The stone walls of the barn are extremely thick and had originally been constructed with a substantial 'lean' to the base of one elevation, which needed to be straightened. Much of the stonework was

"THE GLASS ROOF GIVES US WARMTH AND LIGHT EVEN ON THE WETTEST DAYS, AND IS FANTASTIC ON CLEAR NIGHTS."

Smaller windows and lower ceiling heights create a cosy atmosphere in the sitting room. The rustic fireplace was made from three reclaimed kerb stones.

underpinned and rebuilt by a stonemason, with an internal skin of insulated Thermalite blockwork constructed on new foundations inside the capacious building, and a reinforced concrete ring beam girdling the top of the walls. New Spanish roof slates were purchased for a grand total of £2,000. "They were slightly kinked – which actually gives the roof far more character," says Richard.

Reclaimed bricks were hand dug from a local tip and used for the reveals. Over time Richard had collected a large supply of home grown timber from his tree surgery work for use on the project, which was to have a considerable impact on the end result. Huge pine timbers hold up the glass roof above the dining room, and were also used for the glazed office extension, with green oak chosen for the linear ceiling beams in the kitchen and sitting room.

"THE LEARNING CURVE WAS IMMENSE. I ENDED UP TACKLING EVERYTHING FROM LAYING CONCRETE TO PUTTING IN BEAMS."

Richard helped a man in the local village whose truck had broken down and discovered that he made windows. "One company had quoted us £27,000 for our casements," says Richard, who managed to purchase some hardwood extremely cheaply from a boat builder – and then discovered why. It was so hard as to be almost unworkable. "Chapel Joinery charged £6,000 to make all of our traditional hardwood units, but ended up blunting their tools on the wood in the process!"

The couple set themselves a budget of £70,000 for the conversion,

"IT REALLY IS VERY SATISFYING TO CREATE SOMETHING FROM NOTHING, AND IT HAS SAVED US A GREAT DEAL OF MONEY."

but spent an additional £15,000 on interiors. To keep costs down Richard purchased all materials, managed the subcontractors – paid on a daily rate – and laboured on almost all aspects of the build. "The learning curve was immense," he says. "I ended up tackling everything from laying concrete to putting in beams."

By cutting into the thick walls and removing several tons of stone, Richard was even able to incorporate a larder and extra WC on the ground floor.

Once the house was ready to decorate, Melanie came into her own. Most walls are painted white but, in each room, one wall of strong, vibrant colour prevents the interiors from becoming bland.

The kitchen features both a reconditioned oil-fired Aga and an old eight burner Chester stove, which was salvaged from a Chinese takeaway. Local chestnut was used for the handcrafted kitchen cupboards and the island unit which is inset with marble, with the large Belfast sink costing £50 from an architectural salvage company and supported on a frame constructed from old pallets. "It really is very satisfying to create something from nothing," says Richard, "and it has saved us a great deal of money." ■

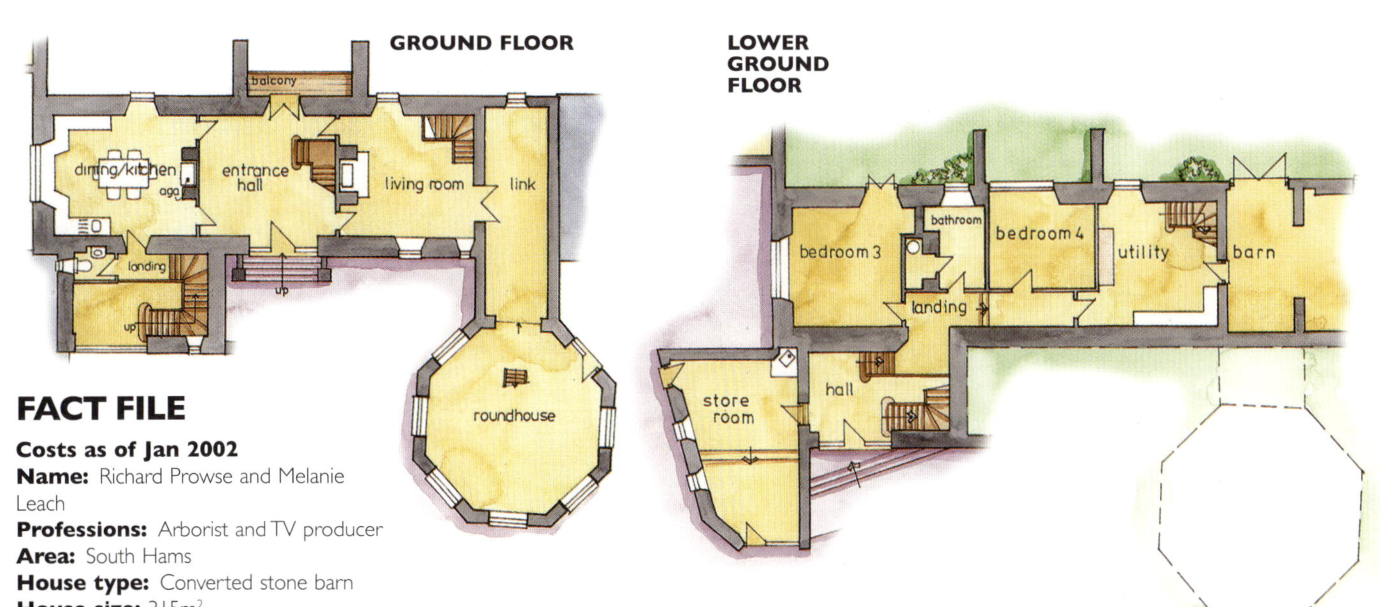

FACT FILE

Costs as of Jan 2002
Name: Richard Prowse and Melanie Leach
Professions: Arborist and TV producer
Area: South Hams
House type: Converted stone barn
House size: 315m²
Build route: Self-build/subcontract
Construction: Blockwork and stone walls, slate and glass roof
Warranty: Structural engineer's certificate
Finance: Bridging loan/mortgage from Northern Rock
Build time: May – Dec '97
Land cost: gift, valued at £70,000
Build cost: £85,000
Total cost: £155,000
Current value: £550,000
cost/m²: £270

51%
COST SAVING

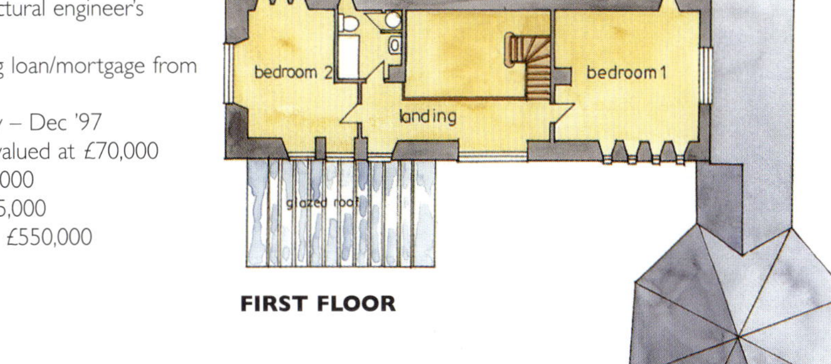

FLOORPLAN

By raising the roof slightly, an additional level was added to the barn. Two bedrooms, a bathroom and utility room at lower ground floor level create a self-contained unit for guests. The ground floor, is accessed from the upper garden and houses the kitchen, sitting room and a galleried dining room roofed in glass. Two further bedrooms, a bathroom and dressing room are situated on the first floor.

USEFUL CONTACTS

Richard Prowse Arborists: 01364 72277 Architect - Peter Thompson Architects: 01202 842266 **Structural Engineer** - David Golightly: 01392 411 733 **Windows** - Chapel Joinery: 01752 698471 **Window Film** - ADS Window Films: 0800 594 2583 **Ironmongery** - Len Wakeham: 01364 72250 **Wooden Flooring** - Mick Corum: 07980 148 266 **Landscaping** - Flete Gardens: 01548 830 435 **Timber Supplies** - ProARB: 07074 007 077 **Plumbing** - Soper Plumbing: 01752 403353

ARTFULLY CONVERTED

Jennifer Newman and Bernard Rimmer have turned an old barn into an elegant minimalist home and gallery space.

WORDS: JASON ORME PHOTOGRAPHY: NIGEL RIGDEN

The unusual steel roof is in fact a Finnish-created form of soft steel that is rolled out on site and then hand-crafted to turn up at the edges.

THERE'S USUALLY A wonderful moment that occurs when visiting one of the growing number of self-styled 'minimalist' houses in this country when the owners, having carefully shown you around the rest of the house with its perfectly manicured detailing and impeccable, almost fascistic tidiness, reluctantly open the doors to a forgotten utility room or storage space and everything – all the clutter of years of living, of old toys, books, clothes – comes crashing out over you.

That this doesn't happen in Jennifer Newman's exceptional Wiltshire barn conversion – created in collaboration with her partner Bernard Rimmer – is at the same time highly admirable and deeply frightening. Even Bernard seems a little perturbed when he opens the push handle doors to the utility space revealing a well-ordered and rather nicely finished sink, washing machine and shelves — with not a towel out of place. And, yes, while it is Jennifer's impeccable attention to detail that has been the driving force behind the project, it becomes quickly apparent that this is a pretty formidable couple for a project of this sort. Jennifer – one of the leading artists in the UK and very much part of the London fashionista – is of course an avowed minimalist but as an artist has an incredible attention to detail and ability to visualise a finished piece, whether it be art or a house, with a self-discipline and rigour and pure and simple creative imagination that would put 99 per cent of the RIBA list to shame. Bernard, however, is the perfect foil; a construction consultant, and while it would be too easy to conclude that he was the practical grounding to some of Jennifer's more left-field ideas, it seems that without his input, this project would never have got off the ground.

And it certainly wouldn't have stayed there. One of Jennifer's grand ideas was that no supports for the roof should be visible in the main space – the problem being that the whole structure would sway in the wind. Cue Bernard and the structural engineer, scratching their heads and looking at

"I WANTED TO PAY HOMAGE TO THE ORIGINAL BUILDING, BUT ALSO CREATE SOMETHING THAT SERVED A PURPOSE FOR BOTH OF US." ➤

The original intention was to leave the flue for the fireplace exposed but Jennifer felt that it would clash with the extraction system she designed for the kitchen space. Instead, an individually designed fireplace, with concrete plinth formed around a plasterboard frame, forms a fine focal point.

A pre-formed concrete floor was chosen as it didn't – unlike stone or wood – create lines to take away from the aesthetic flow of the space. It mirrors the concrete work surface in the kitchen area.

JENNIFER HAS AN EXCEPTIONAL ABILITY TO VISUALISE WHAT SHE WANTS: IN THIS CASE IT WAS HUGE VOLUMES, AN OPEN PLAN ARRANGEMENT AND MINIMALISM.

calculations. The engineer's remarkable solution is the first of its kind in the country — a structure consisting of zigzag style steel bracing which runs the length of the roof and takes the load. The steel enables the large spans that are essential to the success of this remarkable home.

Jennifer spent a couple of years looking for a suitably inspiring project – she was after a pretty cool home that could be combined with fully functioning commercial gallery space in which she could show works by herself and other talented artists (there are a few very tasty pieces by the revered Sandra Blow in the living space). In a delightfully secluded little enclave of Wiltshire, they came across a barn which was a ramshackle arrangement of two buildings clumsily bolted on to each other, dressed conveniently in black but with little more to them than some nice clay pantiles on one of the barns, an appealing courtyard style arrangement and bags of potential.

"I wanted to pay homage to the original building," says Jennifer, "but also create something that served a purpose for both of us in addition to enabling me to express myself in a different format." And boy, when an artist of Jennifer's talent and imagination goes to work on an old barn, it really is something to behold. H&R visits on a rather busy Saturday morning for Jennifer, who is already promoting the gallery's pieces. One gets the distinct sense, however, that it's the finished barn itself that is somehow all part of the show. H&R gives a rather studied "Wow" upon being confronted by the sheer volume of the central space. Later on, as we leave, a rather stuffy architect comes to visit for the exhibition and utters an astonished "Christ!" as he enters. It's entirely thrilling to see the very satisfied look on Jennifer's face – after all, if creating something from scratch having put in the amount of creative energy that Jennifer has done is about anything, it is about enjoying other people seeing it for the first time.

"I interviewed 15 architects initially," says Jennifer, "and it was a rather frustrating process. If their first question was 'What's your budget?' they usually pretty quickly lost interest shortly after I told them what it was. I invited a handful out to see the barn and come up with initial scheme sketches – I'm very much of the opinion that experts in their field should be left to do what they do well – but all seemed to have their hidden agendas. Mainly, they wanted to use the project as a prototype for their own ideas ➤

The two sections of the barn originally abutted each other – but Jennifer and Bernard decided to create a distinct separation between the two, which now forms an informal entrance to the gallery section of the barn.

and prejudged views on what we wanted. It was very disappointing.

"Eventually, I decided to give an architect friend, Max Neal, a call about how to proceed, and he offered to take a look over the drawings for us on an executive level – and I employed a local architect to come up with the plans to get us through planning. I found a young live wire who was working at a local practice renowned locally for its traditional conservation work. When he heard what I wanted he seemed to jump at the chance!"

Jennifer and Bernard managed the project themselves, renting a place nearby and using subcontracted labour. "That for me was the real low point of the project," sighs Jennifer. "We eventually found some really good local people, but it was so deflating when they didn't turn up when they said they were going to and didn't even call. Some just didn't grasp what we were after in terms of attention to detail, but eventually we found the right men."

With a client as demanding as Jennifer it's actually quite surprising that she managed to get anyone to stay. "Everything was agonised over," she says. The attention to detail and the remarkable use of materials on this project is second to none. All the classic minimalist trappings are there – underfloor heating, hidden storage, perfectly white walls, shadow gaps – but this level of perfection can only have been achieved by an artist who spends months agonising over the slightest touch on her abstract paintings. It takes a pretty exceptional eye to pay this much thought to, for instance,

"IF THE ARCHITECT'S FIRST QUESTION WAS 'WHAT'S YOUR BUDGET?' THEY LOST INTEREST WHEN THEY HEARD."

the way the trusses act as a visual divide for the kitchen and living areas of the space; the exact proportions of the kitchen worksurface (more of which later); the configuration of the kitchen unit handles and the appliances ("I would have loved to have designed them all myself as I simply couldn't find what I wanted – for the same reason I had to design some of the furniture and the table and stools in the kitchen," Jennifer explains); the exact and pleasing bulk of the kitchen tap ("A happy mistake"). H&R spends a lot of its time looking through squinted eye at imaginary lines drawn between edges of kitchen worksurfaces and windows. It's all rather overwhelming and very humbling, but it's this kind of it-feels-great-and-looks-even-better-but-you-don't-know-why situation that really sets this house apart from the crowd.

This house is all about light and the windows are again a true one-off. The huge windows and sliding/pivot doors – which cost an eye-watering £40,000 – are framed in galvanized steel, not least because it's only steel that can support this sheer size of glass but also because, according to

The large format pivot doors had to be framed in steel to support the weight of the glazing.

Jennifer, "it fits in with the industrial feel of the barn." There is also a massive amount of concrete – a precast worksurface finished in black which is preformed around the sink and is reassuringly heavy and very chic – and a floor which feels smooth, looks even smoother and yet, in keeping with the rest of the minimalist ethos, feels very inviting.

"It took me a while to be convinced by all this," says Bernard. "Jennifer has an exceptional ability to visualise what she wants and the huge volumes, open plan arrangement and, of course, the attentive minimalism were things that I thought I might have problems with. I was, for instance, delighted to get a rug for the living space, and I feel that of all the wonderful elements, it is the fireplace that is the real success – but having spent a rather peculiar first few weeks living with it, I now love it. I wouldn't dream of living any other way, in fact."

With a perfectly designed landscaping scheme just about to get under way, and with the hard-wired automation systems due to go live when money allows, this looks like being a home that can only improve with time. Jennifer and Bernard have created something quite special here. Lucky visitors to Jennifer's gallery get to see the house for free and it has become a wonderful backdrop to the amazing art. Perhaps, in some eyes, it should be the other way around. ■

GROUND FLOOR

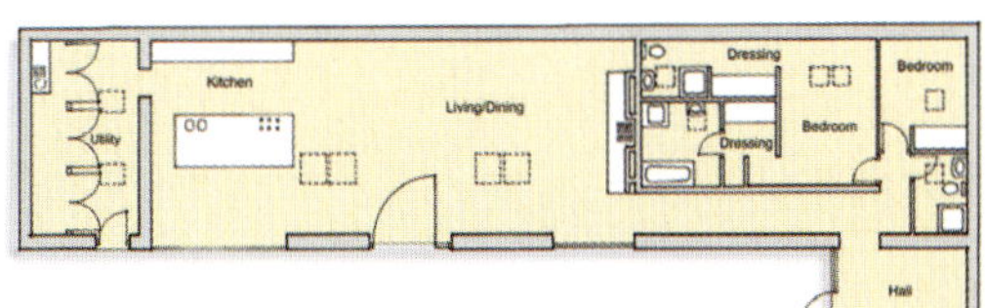

FACT FILE

Names: Jennifer Newman and Bernard Rimmer
Professions: Artist/gallery owner/ furniture designer; construction consultant
Area: Wiltshire
House type: Barn Conversion
House size: 280m²
Build route: Self-managed
Warranty: Architect's Certificate
Finance: Private
Build time: Oct '03 – Oct '04
Barn cost: £195,000
Build cost: £350,000
Total cost: £545,000
House value: £850,000
Cost/m²: £1,250

36%
COST SAVING

FLOORPLAN

The single storey structure is broadly divided into living and working accommodation. The main bedroom has 'his and hers' en suites while the open plan living/dining/ kitchen forms the heart of the house.

METAL ROOFS

Suitable for a wide range of roofing and cladding applications, the Finnish based Rannila Casetti Structural Trays – consisting mainly of galvanized steel – used by Jennifer can be rapidly installed to achieve long spans over nine metres, eliminating secondary steelwork. They can also be installed as a stressed skin configuration to efficiently distribute roof loads through the building frame. Casetti trays are available in an extensive range of depths, coatings and attractive internal finishes. www.rautuk.co.uk

USEFUL CONTACTS

Scheme and planning architects – Simon Morray-Jones: 01225 787900; **Project architects** – Langley Hall Associates: 0118 932 0980; **Consulting structural engineers** – John Tooke & Partners: 020 7820 0297; **Concrete floor specialists** – PB Projects: 01752 294963; **Steel roof** – Ruukki (UK): 01452 421234; **Concrete worktop** – Cast Advanced Concrete: 07979 905667 guy@castadvancedconcrete.com; **Kitchen units** – Trade Kitchens (Dorset): 01747 850990; **Lighting design** – Cook Associates: 01425 674840; **Heating design** – Cudd Bentley: 01344 628821; **Steel windows/sliders/pivot doors** – Faherdex: 01923 247519; **Groundworks, masonry and frame** – Domonic Teversham: 01258 458806; **Partitioning/dry-lining** – Vivid Commercial Interiors: 01305 268944; **Gabion walls, landscaping** – Newhouse Contractors: 01747 823731; **Landscape architect** – Calonder: (+41) 21907 2875 (Switz)

CONVERTING A SMALL STONE BARN ON A BUDGET

GRAIN OF GOLD

Rosie and James Mussen have converted a derelict stone granary on their Devon farm into a stylish cottage with magnificent country views..

WORDS: DEBBIE JEFFERY PHOTOGRAPHY: NIGEL RIGDEN

THE OLD GRANARY must be one of the smallest conversions ever!", laughs James Mussen. "The building is just 30m^2 in size, and it took a great deal of planning to fit in a bedroom, bathroom, kitchen and living and dining areas. It was quite a challenge."

James and his wife, Rosie, purchased Lower Collaton Farm in Blackawton six years ago. Set in ten acres of meadow, the traditional Devon farmhouse dates back to the Domesday Book and earlier, with a working farm next door. Collaton was owned by the mother of King Harold II of England before the Norman Conquest, and was valued at just £50 in the 17th century.

The farmhouse has a walled south facing garden overlooking a quiet valley, with the Mussens' own pedigree Shetland sheep in the fields. In the small woods at the side of the house is the source of the River Gara, which flows down to the nature reserve at in Start Bay, seven miles away.

Apart from the farmhouse itself, which sleeps 13, there are a number of separate outbuildings – including the stables, which the ➤

The Old Granary is tucked in one corner of the grassy courtyard, and boasts views across the surrounding fields. Most recently a pig sty, it is now used for accommodating visiting friends and family and holiday lets.

family have converted into a comfortable reverse level cottage with views from the first floor living area across the fields.

"When we first saw Lower Collaton we fell in love with the house and its location," Rosie Mussen explains. She and James were living in Surrey but had been looking for an old property in Devon which could serve as both a home and business, and the farm had great potential. "We decided to live in the farmhouse and convert the stables immediately in order to generate income from holiday lets," Rosie continues. "For this reason the conversion needed to be completed very quickly, and we employed a team of London builders to undertake the work."

When it came to converting The Old Granary, however, James decided to become more involved and to tackle much of the building work himself.

"IT'S ONLY A SMALL BUILDING BUT NOTHING IS SQUARE... ALL OF THE PURLINS AND RAFTERS NEEDED TO BE REPLACED."

The tiny 30m² building has to make the most of its space – the first floor consists entirely of a kitchen/dining/ living space.

"I enjoy DIY and we have always owned old houses, which means that – by necessity – I've had to be reasonably handy with a hammer and a six inch nail. The granary is attached to an old barn on one side, where I keep bits of wood and tools which might come in handy," he remarks. "Our daughter, Alice [pictured], is very artistic, and undertook the design and project managing. She had the difficult job of fitting a quart into a pint pot, and read a number of books on barn conversions and working with small spaces before we started."

The 18th century stone granary is situated in one corner of the grassy courtyard, and had fallen into disrepair. Planning permission was sought for a change of use for the building, which had most recently been a pig sty. "The farm isn't listed, but the planners imposed certain stipulations regarding the window openings and roof covering," says James, who sketched out the plans himself, which were then redrawn by a draughtsman friend. Reverse level accommodation ensures that the open plan kitchen/living/dining room on the first floor benefits from views across the countryside, with a bedroom and bathroom below.

Corrugated iron covered the roof of the granary, and the planners requested samples of the replacement roofing slate for approval. ➤

Coincidentally, there were a number of small peg slates from outbuildings already on the site which proved ideal. James, who works as a brewery executive, was juggling his day job with converting the granary and renovating the farmhouse, but decided that he could tackle the roofing work himself.

"It's only a small building but nothing is square," he explains. "All of the purlins and rafters needed to be replaced and it was impossible to work in a straight line. Some of the courses of slates had to be staggered to compensate for the shape of the walls, but the building inspector was extremely helpful and gave me some excellent advice."

The walls are built of local shillet and were of sound construction, with a first floor reached by a set of stone steps running up one side of the building. "When the granary was originally built people were far shorter than they are today. In order to achieve enough headroom on both levels we needed to take out the stone setts and dig out the earth floor by an additional 18 inches," says James. "The foundations were pretty minimal, but our building inspector assured us that the structure had been built on stone. I lost a bit of weight using a pickaxe, but I was cracking my head on the doorway – so it was essential work." Laying a new screed gave the Mussens the ideal opportunity to install underfloor heating in the cottage, which eliminated the problem of siting radiators in such a small space.

Underfloor heating has been installed in the granary – beneath reclaimed maple flooring – as radiators would have taken up so much of the limited wall space.

James used a mini digger to create trenches for the services and built a new internal blockwork wall to the rear of the building, which is below ground level. "The granary was very damp, and we solved this by creating a cavity wall inside the stonework, with a trench at the bottom of the cavity designed to drain away the water," he explains. "This meant losing some valuable space, but was vital to prevent water seeping into the cottage."

With one window overlooking the neighbours' sitting area it was important to respect their privacy, and this played a major part in the layout

"HAVING SUCH A FIXED DEADLINE SPURRED US ON, BUT MADE IT A LITTLE MORE STRESSFUL."

A peephole window on the staircase originally served as a ventilation opening for the ground floor prior to lowering the floor level.

of the interior. Single glazed stained softwood windows were made for the existing openings, with a new west facing aperture added in the bedroom. The doorway at the top of the external stone steps has been replaced with glazing, so that access is now at ground floor level only – leading directly into the bedroom which has a contemporary style en suite bathroom.

At the end of the seven month project the couple were just £2,000 over their original budget, and have an extremely useful cottage which may be used by family and friends or let to holidaymakers. "We are also planning to use three double bedrooms in the main farmhouse as B&B accommodation," says Rosie. "It may be a little strange having other families staying in our home, but the money we make from the holiday business allows us to live in this beautiful setting, and having so much space is perfect when our three children want to visit and bring their friends.

"Alice and James started the conversion in October and we needed the cottage to be finished by Easter, as we had already taken bookings for people to come and stay. Having such a fixed deadline spurred us on, but made it a little more stressful. Luckily we finished in time, and everyone who stays here loves The Old Granary – it is such a charming little building." ■

FLOORPLAN

Reverse level accommodation ensures that the open plan kitchen/living /dining room on the first floor benefits from views across the countryside, with a bedroom and bathroom below.

FACT FILE

Names: James & Rosie Mussen
Professions: Brewery executive smallholders and holiday letting business
Area: Devon
House type: One bedroom converted granary
House size: 30m²
Build route: DIY & subcontractors
Construction: Stone, reclaimed slate roof
Finance: Private
Build time: Oct '02 - April '03
Property cost: £10,000
Build cost: £22,000
Total cost: £32,000
House value: £100,000
Cost/m²: £733

68%
COST SAVING

GROUND FLOOR

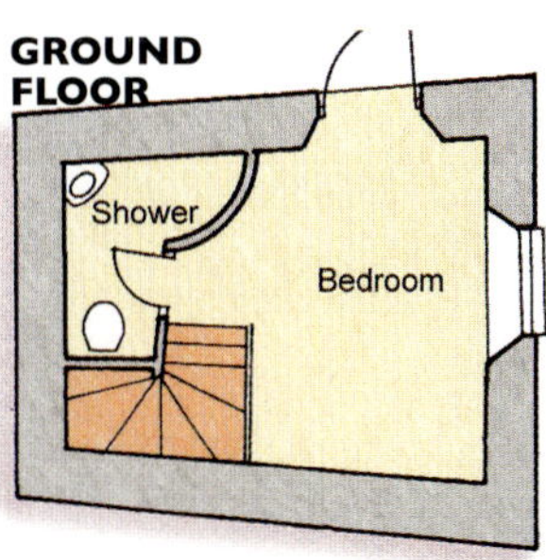

FIRST FLOOR

USEFUL CONTACTS

Holiday accommodation – Lower Collaton Farm: 01803 712260; www.lower-collaton-farm.co.uk; **Interior design** – ACM Interiors: 0208 8888 790; **Building materials** – Travis Perkins: 01604 752424; **Underfloor heating** – Nu-Heat UK Ltd.: 01395 578482; **Plumbers** – Premier Heating: 07855 745040; **Reclaimed maple flooring and oak beams** – Winkleigh Timber: 01837 83573; **Tiles** – Totnes Tile Studio: 01803 865865; **Sanitaryware** – The Blue Tile Co. (Exeter): 01392 439248

CONVERTING A BARN WITH OPEN PLAN INTERIORS

Richard and Rachel McLane have converted a large 18th century stone and brick barn into an open plan home with a contemporary feel.

RUSTIC ELEGANCE

WORDS: HEATHER DIXON PHOTOGRAPHY: JEREMY PHILLIPS

Richard and Rachel made a crucial change to the initial set of plans and decided to make the former granary into a single large kitchen and living space rather than a smaller kitchen and separate garden room.

To bring the old barn up to modern requirements for energy efficiency, the walls were dry-lined with slim, multi-layer reflective insulation and 65mm of Kingspan was packed under the floors.

RICHARD MCLANE BEGAN barn converting a little earlier than most. "My parents live on an old farm and there is a barn tacked onto the end of it, which I probably started converting when I was 14, and worked on till I was about 20," he says. "I nearly got it finished, but didn't quite – my brother took over. I've always liked old stone barns, the features within them, the character and the stories."

When he and his wife, Rachel, saw the cluster of 18th century stone and brick buildings that make up their latest home, the potential was clear. The former granary, milking parlour and livestock barns are tucked away amongst the thatched cottages of an attractive North Yorkshire village, with views over fields.

They already knew the developer who had secured the site, making it easier to negotiate the sale, and there was another big plus: the barns are conveniently close to the Helmsley base of bespoke engineering and staircase specialist Bisca, where Richard is a director – well-placed for keeping an eye on the conversion work in progress.

With so much in its favour, Richard admits he glossed over the site's significant drawback: "There's only a thin layer of soil and then it's bedrock. I had to get the builders to dig out nearly 1,000 tons of rock to get this place into some sort of decent height. To fit the underfloor heating in the kitchen we had to go down a foot to put the slab and insulation in, the pipes, the screed and floor coverings. We had mini diggers in here with peckers on them. The difficulty of ➤

that, the hard labour, is offset by the fact that there's no subsidence on the buildings, because they are on bedrock. Normally farmers just chucked up buildings like this and they are about ready to fall down. Here, there wasn't a crack in sight."

On the downside, there were no services on site. "I knew where I stood, but it wasn't straightforward, because we were digging into bedrock," Richard explains. "Even so, it was only hard work. When you look at the way you are spending money on a job, mechanical diggers are good value and you can see progress."

Although there were already architect's drawings for the conversion, Richard and Rachel, an interior designer, had their own ideas. They were already expecting their son, Dylan, and were keen to create a family home. One important change was to make the granary into a single large kitchen and living space, rather than the smaller kitchen and separate garden room in the first set of plans.

"It's just a better use of space, in my opinion," says Richard. "We've also made a large hall and a utility room, because a house of this size needs somewhere to put stuff. Upstairs, really the challenges were to try and get en suites in without it feeling too poky."

They also spotted an opportunity missed in the original layout. "There was a tin shed lean-to, where the cart would be pulled in, and the architect had just lost that and the building ended short of it. I said to the planners: 'Here's an old farm building and this is a structure that's already here, can I rebuild it?' They said: 'Fine, as long as it's brick, you can use that as planning gain.'" By keeping it, they gained a sun-room extension to the main sitting room; a second, slim offshoot to the side became a useful study.

Naturally, a Bisca staircase was a key feature for the McLanes. Richard and Rachel went for an elegant and subtle, helical design, combining slim steel with sandstone treads. "It's not the most flamboyant one," he says. "I wanted to design something that was soft and timeless. We built the barrel vault out of engineering bricks. We put the staircase framework in first, and gave the builders some templates, and they went ahead and built it."

Turning the focus of the living areas towards the rear made a big difference, with large glazed doors framing the country views. Re-routing the drive around the side of the barns has also increased the sense of privacy.

The village falls within the North York Moors National Park, and the McLanes' revised plans went before its planning authority. "You could argue all day about whether you like their policies, but they

The helical Bisca staircase combines sandstone treads with a slim steel handrail. ➤

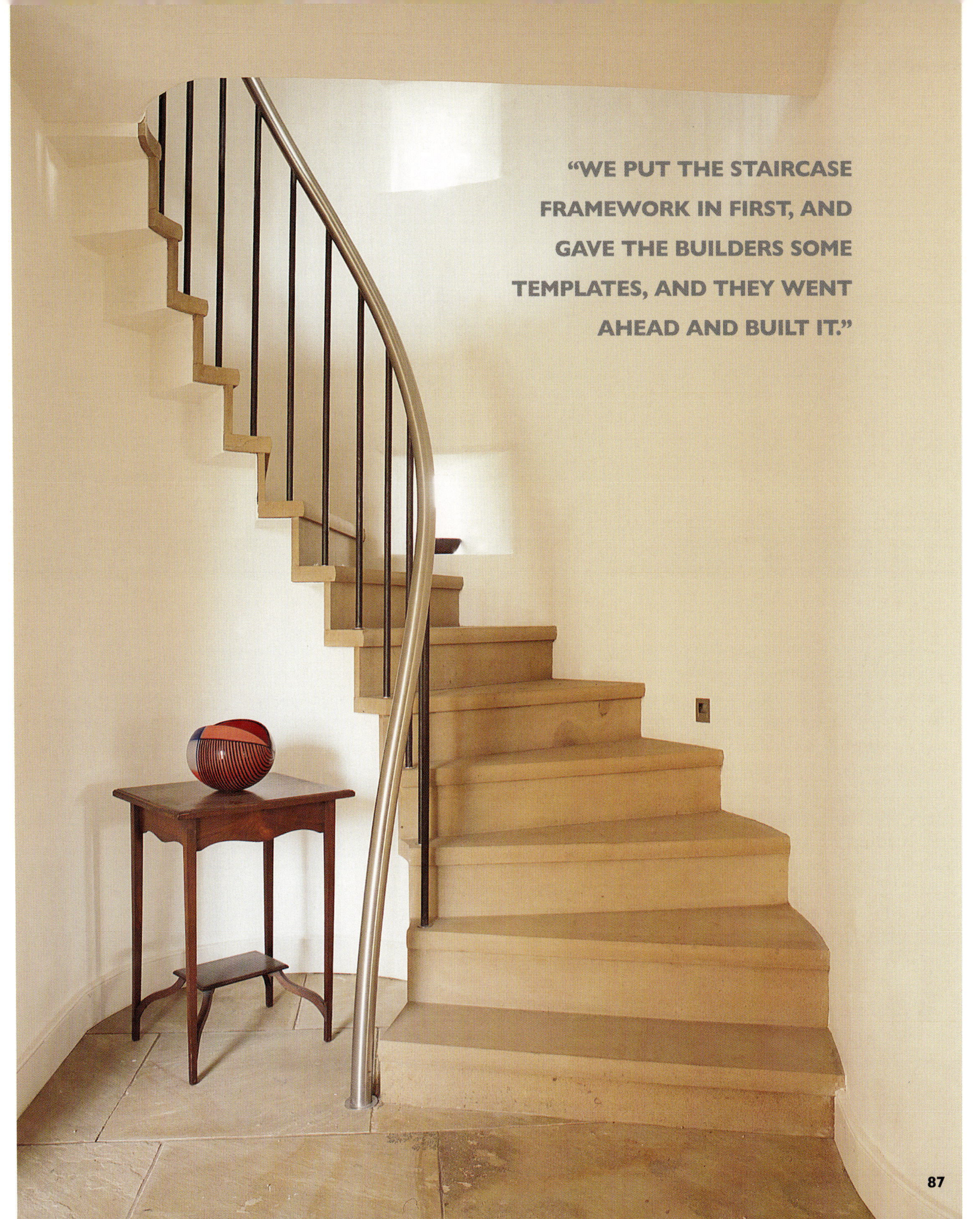

"WE PUT THE STAIRCASE FRAMEWORK IN FIRST, AND GAVE THE BUILDERS SOME TEMPLATES, AND THEY WENT AHEAD AND BUILT IT."

do have a clear direction," says Richard. "If you ask them a question, they give you an answer. My experience of planning in the past is that it's a bit of a black art. 'We will consider it' is the phrase that comes up, whereas the planners on this were very clear on what they did and didn't want. They were very helpful to discuss things with, and it was good to know that we'd got it right, before we put the plans in."

For Richard, the chance to drive a development, without having to refer to clients or designers, was clearly enjoyable. "I love making decisions," he laughs. "If you make a mistake, you can always put it right later, or live with it. It was a pleasure to just say: 'We'll do that.'"

The generously sized hallway was influenced by Richard's childhood home. "My parents' house probably goes back 400 years, but it's mainly a Victorian farmhouse and was in quite a bad condition, and they did a lot of work to it, including putting quite a big hall in. I remember thinking when I was 12 or so, that this was an odd thing to do, but it's actually a really important 'room'. If the large hall wasn't in this barn, if it was narrow and poky and turned off into the dining room, the whole house would have a completely different feel."

He followed his father, too, in choosing underfloor heating. "I remember telling him one thing he could never do was make the house warm, because they never had central heating, and living in an old Yorkshire farmhouse was cold. Sure enough, my father came up trumps and put underfloor heating in throughout the ground floor. That was 20 years ago – he was a pioneer – and it still runs perfectly."

The barn walls were dry-lined with ultra-slim, multi-layer reflective insulation and 65mm of Kingspan was packed under the floors. "I did look at insulations made of timber fibres and sheep's wool, and they just didn't compare. We were looking at far more thickness to get the effect we could get with the high-tech solution and it was just going to bring the whole place in. Although we're not restricted for space in the main rooms, in the bathroom and the landing corridor I just felt it wasn't worth pulling the space in."

There was one late-stage change to redraw the main and master en suite bathrooms. "The en suite is quite big with a shower now, but it was going to be the main bathroom and a bigger, more grand affair. I thought there was no need for both to have a really smart en suite and a big bathroom."

Building progress proved speedy and efficient. "The builder we used was just able to control labour," says Richard. "He was very good at throwing men at the job and making them work in their right areas, whereas I do think if you have a building firm that has only got five men and they are trying to keep another job down the road going too, it will slow things. At some stages there were 25 men in here: people outside on the roof, people inside. It was buzzing."

WE'VE LEARNED A LOT HERE, AND WE'RE REALLY PLEASED WITH THE WAY IT'S TURNED OUT."

Rather than stain or distress the new timbers to blend with the originals, they chose to leave them in their natural state. "The point was discussed, but I thought: 'Why?' It's rough-sawn timber, you can see that in the saw marks. It's never going to look quite right, it will always look like a repair job, so why not show it as it is?"

Against the dramatic backdrop of the original A-frames and exposed beams, the McLanes have kept the interior decoration simple, with a neutral colour scheme and an emphasis on natural materials, mixing traditional features like the reclaimed stone fireplace and contemporary details such as the bespoke door furniture in stainless steel.

They had a set budget, but it was inevitably stretched as they pursued quality finishes, including the oak kitchen furniture and oak windows. "We just kept on spending," Richard laughs. "For a quality job that is well crafted, I don't really quibble. For items like stone flags, where a supplier is just selling them on, I'd go for the cheapest, but if someone is reliably producing good craftsmanship, I wouldn't question their costs. If I couldn't afford it, then I would have to look elsewhere for something I could afford."

Moving in was delayed slightly, but Richard, Rachel and Dylan were able to settle in before Christmas. "The windows arrived late in the job; the heating wasn't on as soon as it could have been. We could have dried the place out a bit more. You don't quite appreciate the fact that as it has never been a house, how damp it naturally is."

It's tempting to stay, but they admit that, though they love the results, they are hooked on developing and are already thinking of the next challenge. Even so, they're convinced the choices they made for their barn were the right ones for the long term. "I'd like to think that anybody who buys this house in the future is not going to rip the staircase out," Richard says. "It works. They're not going to pull all the windows out either. I don't think many people would want to change the core of the house and the layout.

"My next house might be more conventional in its groundworks, but we've learned a lot here, and we're really pleased with the way it's turned out." ■

FACT FILE

Names: Richard and Rachel McLane
Professions: Director of Bisca and interior designer
Area: North Yorkshire
House type: Converted barn
House size: 250m^2
Build route: Self-managed with subcontractors
Construction: Yorkshire stone and brick with pantile roof
Finance: Private
Build time: March '04 - Oct '04
Land cost: £180,000
Build cost: £200,000
Total cost: £380,000
House value: £625,000
Cost/m^2: £800

39%
COST SAVING

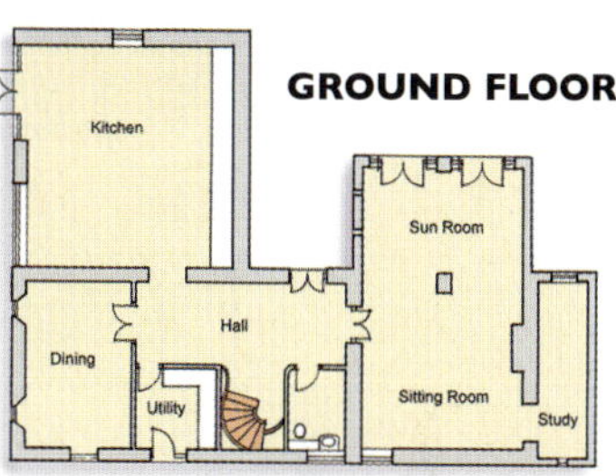

FLOORPLAN

The combined sitting and sun room are the focal point of the ground floor layout, while upstairs four bedrooms and three bathrooms complete the accommodation.

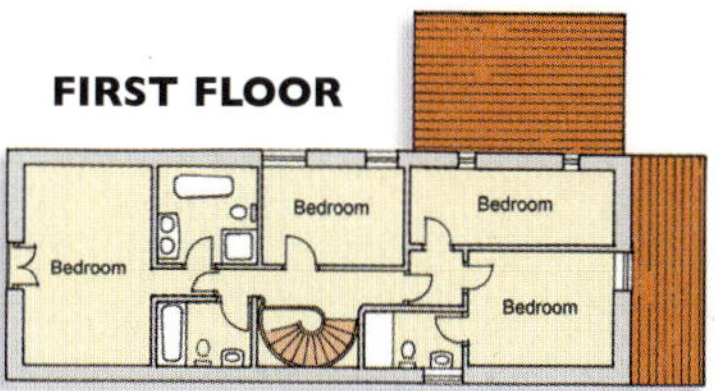

WORKING WITH OLD BEAMS

A gentle approach is the key thing when it comes to cleaning up the old trusses, posts and beams found in old barns. Avoid pressure washing and use gentle non-toxic cleansers – or even soap and water – to remove superficial stains, but any deep staining will have to be cut out. Use an oil such as linseed or beeswax to treat after cleansing.

USEFUL CONTACTS

Staircase, lights, dining table and design – Bisca: 01439 771702 www.bisca.co.uk; **Doors, windows, kitchen door fronts and benches** – S Taylor & Son: 01751 472143; **Garden** – The Landscape Practice: 01609 772554; **Ironmongery** – Ryelund: 01439 772802; **Plumbing** – Plumbing Imports 0845 310 80 59 www.plumbingimports.co.uk; **Green oak** – Walls Timber Yard: 01439 788554; **Oak flooring** – Duffield Timber: 01765 640564 www.duffieldtimber.com; **Reclaimed fireplace stone** – Drings: 01751 417237; **Stone flooring** – North West Reclamation 01282 603108; **Stone treads for staircase** – F Jones (Cleveland) Marble & Granite: 01642 241195; **Underfloor heating** – John Guest Speed Fit: 01895 449233; **Bathroom sandstone tiles** – Topps Tiles: 0800 783 6262 www.toppstiles.co.uk

RUSTIC CHARM

David and Phillipa Johnson have converted a ramshackle old barn into a family home and crowned the project with a large garden room extension.

WORDS: CLIVE FEWINS
PHOTOGRAPHY: ROB JUDGES

Oak floors upstairs and down, and also oak doors, several of them glazed, complete the look. The fireplace in the sitting room, though it could be taken for New England in style, is in fact an Arts & Crafts design, made a few miles away in the Cotswolds.

DAVID AND PHILLIPA Johnson drove half a mile down a farm track in rural Oxfordshire in December 1995, expecting to find a part-finished barn conversion at the end. Instead they found themselves confronted with what they nowadays jokingly call "three and a half walls and a roof."

Disillusioned, they beat a hasty retreat. It was only when they looked hard at the price – £98,000 – and happened upon their would-be neighbours in a local pub that they decided to pay a second visit.

"We really had no intention of doing a project of this sort," says David. "However, when we thought about it a little more and considered the rural location and the size of the plot that came with the barn – half an acre – we decided to take another look."

The barn – part of a mid-19th century former farm complex – was built as a cart shed beneath and grain store above. It had a slate roof, sound stone walls, and softwood beams, most of which were capable of being retained and used as ceiling features. Phillipa, who at the time worked as a retail buyer in London, specialising in New England and Shaker-style furniture, could see the potential. David, an accountant, had a sneaking feeling (that proved correct) that property prices were about to take off again in a big way.

Through a contact in nearby Abingdon, where he was working at the time, David met architect Alan Drury. Alan relished the prospect of a barn conversion, and as the Johnsons had by then realised that all the

"IF WE HAD NOT ASKED [THE BUILDERS] TO LEAVE THE SITE [NEAR THE END OF THE PROJECT] THEY WOULD PROBABLY HAVE WALKED OUT..." ➤

more tempting-looking barns for conversion were out of their price range, he was asked to come up with some ideas.

David and Phillipa bought the plot and the building – along with a large pile of stone with which they intended to build the front wall – and a set of plans which came with the property. The first thing Alan did was to reject these plans and suggest that they would be better off using the stone in the very big garden they planned to have.

"Alan fully understood the brief when we said we wanted clean, simple interior lines and for the house not to be too 'cottagey'," says David. "At the same time we wanted to 'go with' the building – we didn't want an interior that bore no relation to the original, as so many barn conversions seem to end up. We also wanted to reuse as many of the existing materials as possible."

Alan's scheme introduced light through the south-west-facing rear via some carefully positioned windows that were not so large that they ruined the proportions of the original. He also included a floor-level exterior deck, so they could make the most of this aspect, which overlooks woodland.

"The previous plans had been drawn

The garden room extension, completed some seven years after the barn conversion, provides the final accommodation requirements for the Johnsons and their young daughter, and adds an elegant design touch to the original building. ➤

The barn – part of a mid-19th century former farm complex – was built as a cart shed beneath and grain store above.

"ALAN REDESIGNED THE STAIRS AND MANAGED TO FIND US AN AMAZING AMOUNT OF STORAGE CONSIDERING WE HAVE NO LOFT, AND CREATED THE SORT OF FREE DOWNSTAIRS CIRCULATION WE DESIRED."

up without a great deal of thought," admits David. "In particular the plans for the third storey – we have two rooms in the roof – would probably not have passed building regulations. Alan redesigned the stairs and managed to find us an amazing amount of storage considering we have no loft, and created the sort of free downstairs circulation we desired, with a central chimney stack. He achieved a light, airy, contemporary feel, but also managed to retain the ceiling joists and roof timbers. The joists would have completely disappeared had we followed the scheme that came with the barn. That would have been a great shame. As it is the ceilings suit the look we wanted – as much timber in evidence as possible."

The most dramatic feature of the original section of the house is undoubtedly the cedar-clad frontage which gives its name to the house – Cedar Barn. The vertical cladding and the very asymmetrical look that Alan has given to the front elevation is in stark contrast to the other converted farm buildings in the complex.

"This building was always different from the others in the terrace because they have stone frontages while this one had a rendered timber frame," Alan explains. His argument for a timber framed frontage to the house rested on this. "Why do something different and add stone when there is already a perfectly acceptable solution staring you in the face?" he says. "By sticking to a timber frame and inserting new oak posts in exactly the positions where the columns were in the frontage before, we have echoed the original and at the same time created something 'different', which is what the Johnsons wanted."

With only minor hiccups, work kept to schedule and the Johnsons moved in during December 1997, six months after the builders had started. However, there followed a series of small disputes with the builders, who, according to David, were by then bogged down in a distant large contract. "As far as I could tell a large contract had gone wrong and they decided to go into voluntary receivership," David says. "The result was that all the final jobs on this contract, including the final snagging, dragged on. By October 1998 we were still dissatisfied and withheld some of the payments. We were able to prove to them that they had miscalculated quantities and over-ordered, and we parted company. If we had not asked them to leave the site at this point they would probably have walked out. We finally paid off the builders in January 1999. In the meantime we used the builders who had originally been our second choice – Avon Construction – to complete the unfinished work. It took them about three weeks.

"Throughout this period, Alan, who had supervised the job throughout, helped us to resolve all these issues. As we really knew nothing about building, to my mind Alan represented the best use of our money, especially in view of all the problems we encountered towards the end. Without him we would have been really stuck."

Three years later, and with the birth of their daughter Stella, now three, in between, the Johnsons started again. They had no hesitation in going back to Alan to design them a large single storey garden room ➤

"AS WE REALLY KNEW NOTHING ABOUT BUILDING, TO MY MIND, AN ARCHITECT REPRESENTED THE BEST USE OF OUR MONEY. WITHOUT HIM WE WOULD HAVE BEEN STUCK."

that sits at the eastern end of the house and, with its large overhang, presents a striking feature. Once again, wood predominates: the structure, which leads out to a stepped deck, is glazed on three sides and has a Welsh slate roof to match that of the house. The roof pitch is steeper and at a lower level. It has provided additional dining space for year-round use as well as extra play space for Stella.

"It is a lovely room and we have used it a great deal," says David. "The slate roof means it is very definitely a garden room and not a conservatory. However, it is a bit of a hybrid from the point of view of building regulations. It fits in with the requirements of a conservatory in that it is isolated from the rest of the house, but not with the other requirements because it does not have a glass roof. Alan had to argue the case very hard and show that the roof is in fact better insulated than a normal glass conservatory roof. Once this was sorted out the building went up extremely quickly."

Now, after seven years, the entire project is complete and the Johnsons' lovely garden, with the dovecote from the farm complex at the bottom, is looking very mature. "As far as we are concerned we have no desire to move," says Phillipa. "So much of us has gone into the project that it would have to be a very special house to drag us away." ■

FLOORPLAN

The main ground floor living space has been augmented by a side extension to create a garden room. The main bedroom on the first floor has an impressive double height space.

GROUND FLOOR

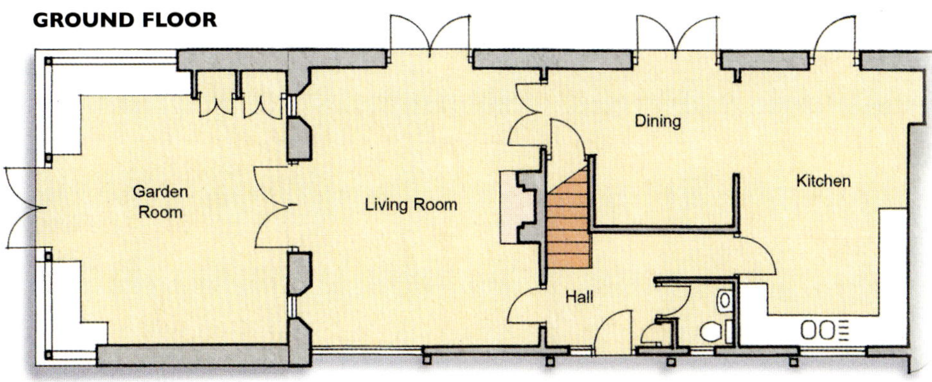

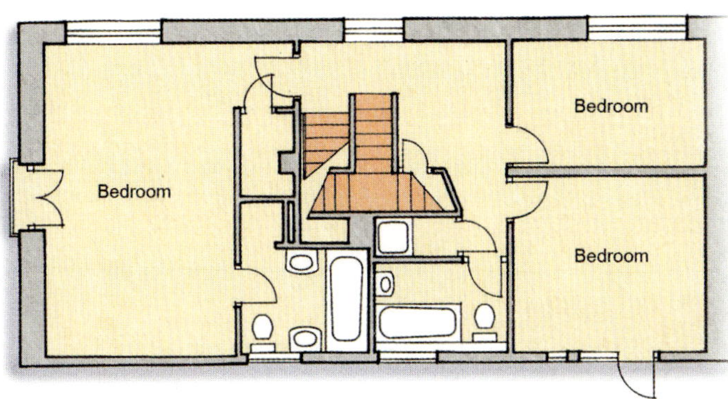

FIRST FLOOR

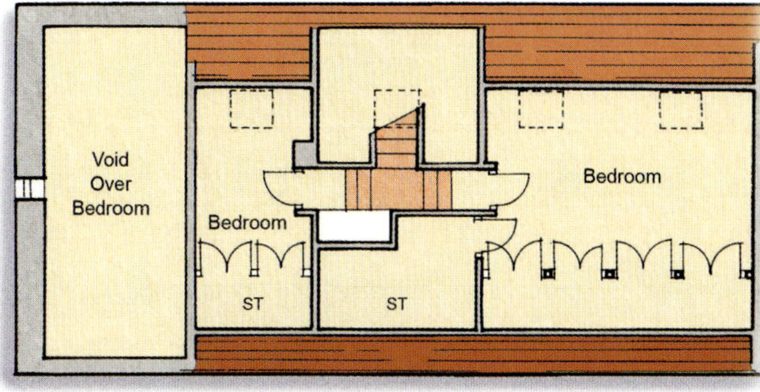

SECOND FLOOR

FACT FILE

Names: David and Phillipa Johnson
Professions: Accountant and homemaker
Area: Oxfordshire
House type: Five bedroom detached
House size: 161m², plus 28m² garden room
build route: Main contractors
Construction: Timber frame at front, solid stone walls elsewhere
Finance: Private plus temporary mortgage from Cheltenham & Gloucester
Build time: Main works June '97 – Dec '97 Garden room: Dec '03 – April '04
Land cost: £98,000
Build cost: £235,000
Total cost: £333,000
House value: £725,000
Cost/m²: £1,237

54%
COST SAVING

Cost Breakdown:

Main contractor	£105,000
Kitchen	£17,000
Lighting	6,000
Second contractor	£5,000
Miscellaneous (including fees) £17,000	
Garden room	£85,000
TOTAL	**£235,000**

THREE STOREY LIVING

All habitable rooms in a dwelling that are more than 4.5m above ground level are considered to be too high for people to safely make their own escape from windows in the event of a fire. The Building Regulations therefore require an alternative means of escape. The usual solution is to include a protected escape route down through the house and to an exit door, formed with half-hour fire-resisting construction and fire-resisting self-closing doors. It is also necessary for all habitable rooms to have an escape window with a clear space measuring not less than 450mm x 450mm. Visit www.communities.gov.uk.

USEFUL CONTACTS

Architect Alan Drury: 01235 553073 **Main contractor house (second contract)** Avon Construction: 01666 505355 **Garden room** Constructive Design: 01235 848649 **Kitchen** Duns Tew Kitchens: 01869 340429 **Lighting** Optelma: 020 7831 2020 **Fireplace** Farmington: 0800 731 0071 **Windows** Fairoak: 01276 38606 **Rooflights** Velux: 01592 772211

Stephen decided to arrange an on site meeting for the planners to give approval to a selection of rooftiles. "Out of 29 samples they only found two acceptable," says Stephen.

CONVERTING A GRADE 11 LISTED BARN

TASTEFULLY TRANSFORMED

Stephen Lewis has converted part of a 16th century Grade II Listed stone barn into a home with dramatic spaces – despite problems with the planners.

WORDS: BILL LAWS PHOTOGRAPHY: COLIN BARRATT

On one side of the hallway is the farmhouse kitchen and on the other the utility room and two large bedrooms, both en suite.

UPON RETURNING FROM New Zealand it took Stephen Lewis several months to find the kind of traditional English property he was looking for. When he spotted-a 16th century, Grade II Listed barn on the Gloucestershire and Herefordshire borders, Stephen immediately recognised the potential lurking beneath the vegetation and corrugated tin roof. "It was the most magnificent barn I'd ever seen – the profile of the building; the stonework; the courtyard and its private location meant it would have been a privilege to own."

The barn came complete with planning permission for conversion into two houses. However his budget was nowhere near enough to finish two properties at the same time and instead he decided to approach them one at a time.

Eager to get friends involved as much as possible in the project, Stephen decided to host a series of 'barn storming' parties. Despite the lack of an actual house, his friends arrived with a good supply of wine and plenty of sleeping bags and set to on a series of working weekends.

Three years later, upon completion of the first house, he repeated the reunion. This time his friends could admire the picture of a perfect conversion: the circular garden courtyard, overlooked by large, stand-alone, timber framed conservatory, leads through the green oak and glass frontage, into a great hall open to the rafters. By placing the main living area on the first floor and insulating between and over the rafters, Stephen has been able to make use of the roof structure as a feature. He has done the same in the master bedroom on the other side of the hall/ stairwell void.

"THE AUTHORITY OPPOSED SOLAR PANELS IN THE ROOF, WHICH IS A PITY BEARING IN MIND THE NEED TO CONSERVE ENERGY."

The conversion was a less than straightforward journey, however. When building began on stage one, Stephen ordered building materials from local suppliers, reclaiming the VAT later. He teamed up with architect Paul Brice and builder Ian Colwell from Complete Build.

"You can't overstate the importance of having an architect with enthusiasm and love for an old building like this. We changed the ➤

Stephen was keen to preserve as much of the original building as possible while at the same time create a home suitable for modern living. "The existing openings don't let in enough light and you need to be more creative with door openings and glazing wherever possible," says Stephen, who has also preserved some of the original timber wattle.

original plans for the conversion to better meet my needs and it's vital to have drawings which get down to the finest details so the builder knows exactly where he is," says Stephen.

The first part of the project was the construction of an oak framed conservatory in the courtyard. Stephen used the conservatory as bedroom, kitchen and office in the months ahead. One of his regular office duties was negotiating with the planners.

There was a struggle with the planners over a row of Velux windows in the roof – but that struggle paled in comparison with the long negotiations over the roofing material. "The authority opposed solar panels in the roof which is a pity, bearing in mind the need to conserve energy – and the choice of rooftiles was only settled after I arranged an on-site display of 29 different samples from seven different manufacturers. They found only two acceptable. The problem seemed to be that while the local authority knew what was unacceptable, they were reluctant to suggest an alternative."

"THE OPEN PLAN LAYOUT MEANS THAT EVEN SMALL SOUNDS CARRY FAR."

Nevertheless Stephen has successfully converted a 300 year old barn into a 21st century home, without losing the integrity of the building. The rough rendered wall finish, a preserved section of timber wattle, the open beams and some fine stonework all help.

Given another chance Stephen might have done some things differently. "It would have been better to place the oil-fired boiler in the middle of the building, rather than in the utility area at the far end, because it takes a while for the hot water to reach the kitchen taps – although this is remedied by a pump. In addition, the open plan layout means that even small sounds carry far.

"I still consider this a work in progress," Stephen explains, as he completes the second stage. "This is more than just a home – it's a long term lifestyle. I plan to live here for the rest of my life and want it to be perfect." ■

The kitchen, which has Indian flagstones and underfloor heating, came from IKEA.

The living room, open to the monumental oak beams, is situated on the first floor, above the kitchen.

FACT FILE

Name: Stephen Lewis
Profession: Software designer
Area: Glocs/Herefordshire borders
House type: Barn conversion
House size: 300m^2
Build route: Architect & contractor
Construction: Stone
Finance: Private
Build time: Aug '98 – July '01
Barn cost: £75,000
Build cost: £200,000
Total cost: £275,000
House value: £425,000
Cost/m^2: £916

36%
COST SAVING

FLOORPLAN

The hallway has been left open all the way to the ridge to retain the barn's character and sense of space. The kitchen and guest bedrooms are on the ground floor, leaving the first floor available for the main living area and master bedroom, both of which make use of the roof structure as a feature.

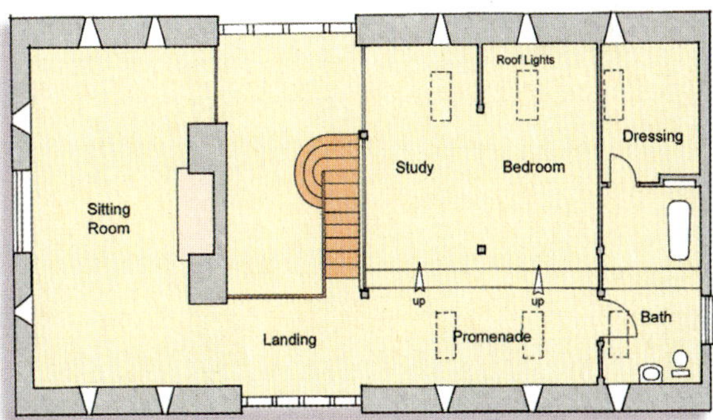

GROUND FLOOR

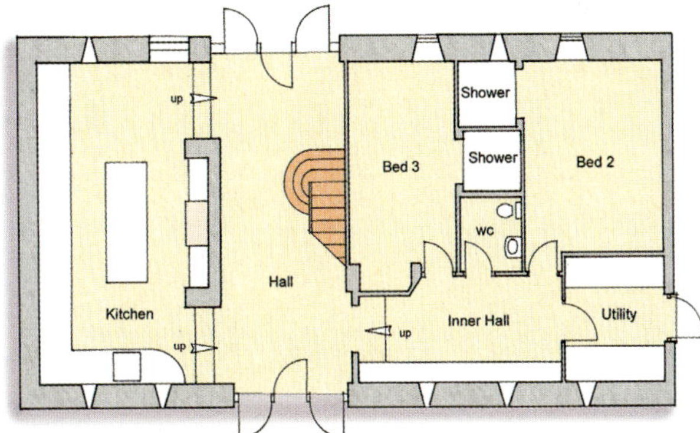

FIRST FLOOR

USEFUL CONTACTS

Architect – Paul Brice, 01600 715810; **Builder** – Ian Colwell, Complete Build 07778 890225; **Specialist joinery** – Mike Griffiths 01531 660580; **Flagstone Floor** – Cardiff Reclamation:02920 458995; **Stairs and joinery** – Charles Davis Joinery:01600 740633

CONVERTING A GRADE II LISTED BARN

ONCE A LOWLY CATTLE SHED

WORDS: HEATHER DIXON PHOTOGRAPHY: DAVE BURTON

Sue and Richard Jobson have converted a 200-year-old Grade II listed stone barn from a dirty, empty shell, into a spacious, bright family home.

IT WAS A grey, wet and windy day when Sue and Richard Jobson first saw the tumbledown old barn in West Yorkshire. But even the bleak moorland weather didn't put them off the property – they could already visualise it as their next home.

The front elevation was unstable, the roof was in a terrible state of repair and the land was piled high against the building at the back. Inside it was a "big, dark, dirty shell, ➤

A chimney, made from reclaimed bricks, was specially built to house the new Aga. Sue and Richard love the open plan arrangement of the kitchen – they had originally planned a separate dining room, but decided they would not use it, so created more space in the kitchen for a table and chairs.

with cattle stalls still standing in rows," but Sue and Richard loved it.

"We knew it would be hard work to turn it into a home, but we could imagine a house with large, spacious rooms," says Sue. "We had been looking for land to build our own house on, but there was nothing we really liked – then this barn came up for auction. It would be very different to our last place, which was a modern detached house on a small development of new builds, but we needed more space as the children were growing up."

Throwing caution to the bitterly cold wind, the Jobsons agreed to go for it, and attended the auction – armed with a ceiling price of £130,000.

"The room was packed, but it soon became clear that there were only two of us seriously bidding," remembers Sue. "It was really nerve-racking. The other bidder offered £125,000. I dug Richard in the ribs and we offered another £2,000. My heart was pounding. It was a huge relief when the hammer came down, despite the amount of work ahead of us."

The building is Grade II listed, so the Jobsons couldn't change the external layout or the windows. With the help of Howarth architect Andrew Jones, they drew up the design for the internal space and submitted their plans to Calderdale's planning department to include an open plan kitchen and sitting area, five bedrooms, five bathrooms and two attic rooms.

"AFTER THE PLASTERING STARTED, THERE WERE TERRIBLE STORMS AND WATER STARTED GUSHING INTO THE LOWER SITTING ROOM..." ➤

The building is Grade II listed so the Jobsons could not change the exterior or the windows.

"WE HAD A PROBLEM WITH THE ROOF HEIGHT WHEN SOMEONE COMPLAINED TO THE LOCAL COUNCIL THAT THE PITCH WAS HIGHER THAN IT SHOULD HAVE BEEN."

The first job was to clear the land with two JCBs, to create space for a courtyard and detached double garage. A mini digger was hired to reduce the internal floors from four levels, stepped up the hillside, to just two.

Builder Russell Wilkinson of Rose Construction tanked the walls – using masonry bricks and Sika rendering – at the gable end to prevent water seeping in as it ran down the hill at the back of the barn.

The floors were dug down by two feet so a steel mesh could be laid ready for the concrete floor, the internal walls were breeze-blocked and the external walls insulated with thermal block. The roof came off and the front elevation was dropped and rebuilt, using the same stone. Steel RSJs were craned in and thick steel pillars embedded into the concrete floor to support them, so the property could be raised to roof level. New open roof trusses were used to create attic space, Rockwool and Kingspan insulation installed between the trusses and plasterboard, and a slate roof finally went

The staircase was originally positioned against the sitting room wall, but Sue changed the design to the middle of the room to create more impact. The fireplace was made to Sue and Richard's own design.

on in August 2002 – though not before a five-month setback.

"We had a problem with the roof height when someone complained to the local council that the pitch was higher than it should have been," says Sue. "As soon as the complaint was lodged we had to stop building work until it was sorted out. It took months but in the end we were found to be within the specifications of the drawings and allowed to continue. It cost us time and expense we could have done without. It was a worrying time."

During this time Sue and Richard were living in their old house, which they had sold. The buyer pulled out due to the delays and bought their neighbour's house instead. "It worked in our favour because we finally sold it for £100,000 more," says Sue. "It also gave us time to work out what we wanted from the new house, so we made a few changes to the layout at this point. These included redesigning the staircase, and swapping the dining room with the study."

When work resumed, the roof was completed and handmade hardwood framed windows and woodwork – including architraves, doors and skirtings – installed, whilst final fix electrics and plumbing were completed.

"Fortunately, all the main services were provided, including gas, because there are neighbouring properties and a village nearby," explains Sue. "The water pressure here is so high that we didn't need ➤

"I KNOW FROM PEOPLE'S REACTIONS WHEN THEY WALK IN AND SAY 'WOW' THAT IT'S A SUCCESS."

power showers and we installed a Keston C55 condensing boiler to heat the house."

A false stone chimney breast was built in the kitchen and another, with a flue, built in the sitting room, then the plasterers moved in to 'pull the whole house together'. Then disaster struck again.

"We thought everything was going well when, shortly after the plastering started, there were terrible storms and water started gushing into the lower sitting room through a storm pipe, flooding the room.

"It was heartbreaking. The floor had to be ripped up and a waterproof Sika render applied, to a metre high all round the room, to prevent it happening again."

The Jobsons decided to turn this room – originally a formal sitting room – into a games room.

In spite of the delays and upheavals, Sue and Richard still maintain that it has all been worthwhile. "I know from people's reactions when they walk in and say 'wow' that it's a success," says Sue. "Even Richard's brother, who thought we were mad to take it on, has admitted that it's worked really well.

"If we did it again we wouldn't expect everything to run smoothly," says Sue. "When you are renovating there are always unforeseen problems. You just have to keep your eye on the goal and remember that, sooner or later, you will achieve the house you dreamt about when you first saw the property." ■

FLOORPLAN

The ground floor has an open plan feel with an informal living area off the kitchen and another off the entrance hall. The barn now includes six bedrooms over two more levels.

SECOND FLOOR

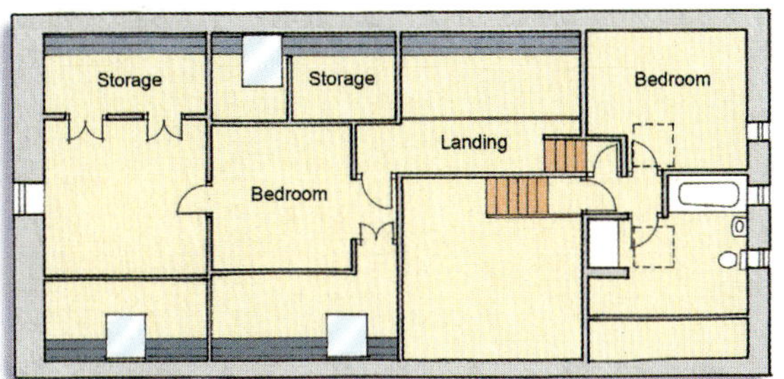

FIRST FLOOR

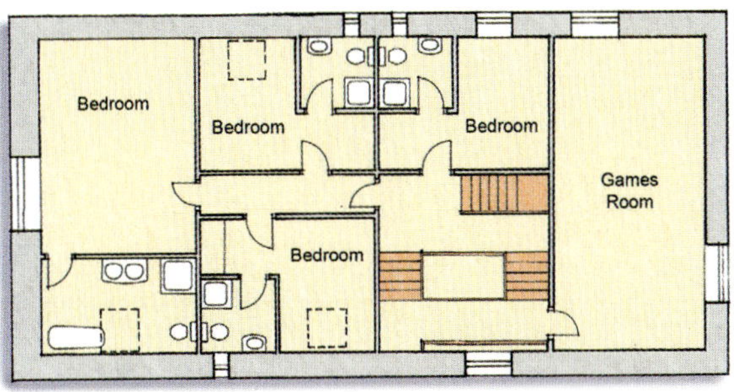

GROUND FLOOR

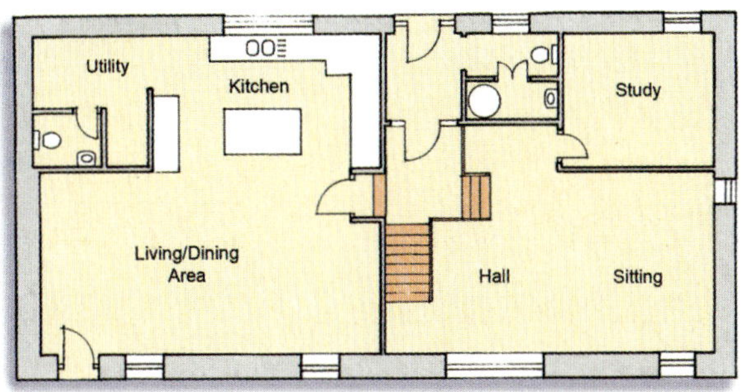

FACT FILE

Names: Sue and Richard Jobson
Professions: Homemaker and former professional footballer
Area: Halifax, West Yorkshire
House type: 200-year-old, Grade II listed barn conversion within an acre of land
House size: 325m²
Build route: Main contractor
Construction: Stone
Finance: Private
Build time: Oct '00 – Feb '03
Land cost: £127,000
Build cost: £450,000
Total cost: £577,000
House value: £850,000+
Cost/m²: £1,384

33%
COST SAVING

Cost Breakdown (selected):

Bathrooms (suites only)	£12,000
Kitchen	£25,000
Sika tanking	£5,000
Aga	£7,000
Internal doors £400 per door (total 22 doors)	
Joinery	£90,000
Kitchen floor	£2,273

WATERPROOFING

Waterproof renders are an increasingly common solution against the problem of water penetration and can be used internally as well as on the outside of the house. Sika, which the Jobsons used internally, is a rapid-setting leak-stopping liquid which is mixed with ordinary Portland cement to produce a paste for rapid leak sealing against high-pressure water infiltration – although it is far from the only brand on the market. The key to the success of any internal waterproof render is to ensure an impenetrable seal at the wall/floor junction.

USEFUL CONTACTS

Windows and joinery Bradmill Joiner in Triangle: 01422 316336 **Kitchen** Little London Kitchen Company: 0113 250 0169 **Verdi Ubatuba granite worktops** Pisani in Matlock: 0845 130 7733 **Stone floor in the kitchen** Lapicida: 01423 560262 **Ceramic floor tiles** Dickies Tiles, Halifax: 01422 320032 **Bath** Keighley Plumbers: 01422 363589 **Stone for doors and windows** Philip Cooper Builders' Merchants, Elland: 01422 377359 **Steel** Walker Metalwork, Elland: 01422 310011 **Plumbing** Nigel Kyte: 01422 344844 **Feature beams** Aaron Beam Centre, Halifax: 01422 300300 **Aga** Bower & Child, Huddersfield: 01484 425416 **Paving** Northern Stone Paving Company: 01274 722000 **Glass** Oddyssey Glass: 01422 359028

CONVERTING A BARN IN CONTEMPORARY STYLE

DISPLAY CASE

Arthur Smart has created an elegant minimalist style home for his parents by converting and extending a barn.

WORDS: HEATHER DIXON
PHOTOGRAPHY: DAVE BURTON

YORKSHIRE-BASED STUDENT ARCHITECT Arthur Smart couldn't have chosen better 'clients' for his first major design project – his own parents. Anne and Tony Smart wanted to convert a former cow barn into an open plan living space which would double as a gallery for Anne's works of abstract art and a place for residential courses.

Arthur's challenge was to transform the long brick building into a modern home which not only provided the right balance and proportions for Anne's huge canvases, but also respected the integrity of the 150-year-old barn.

"The artwork was always key to the design," says Arthur. "We wanted to introduce lots of glass into the place so the building could work as a sponge, soaking up colour and light. The aim was to create a vehicle for what was happening inside and outside the home."

"WE WANTED TO INTRODUCE LOTS OF GLASS INTO THE PLACE SO THE BUILDING COULD WORK AS A SPONGE, SOAKING UP COLOUR AND LIGHT.

Arthur also had to take into account his parents' desire to simplify their lifestyle, getting rid of clutter accumulated over 16 years of living in two other properties on the same site. "They were moving from 250m^2 into something almost half the size," says Arthur. "Space was at a premium and it had to be used well."

Anne and Tony trusted their son implicitly. He had demonstrated his flair for design by removing floors and creating voids in their last house, and they were prepared to let him flex his talents on their ambitious new project, while they kept an eye on its development on a daily basis.

They were also prepared to wait to achieve their ideals – which was perhaps just as well. It took two years to get planning permission to turn the outbuilding into a dwelling with an extension for a studio/workshop.

"The original extension was going to be much longer, but our plans were rejected because it would have almost doubled the size of the building and they wanted us to work as closely to the original footprint as possible," says Arthur. "We turned the delay to our advantage because it gave us more time to sit down as a family and discuss the project in much more detail. It gave me chance to find out exactly what they wanted from the house and how they wanted to live."

The main kitchen units were positioned at right angles to the rest of the layout to minimise the long, narrow effect of the barn.

The original A-frames were treated for woodworm but left unsanded.

Four Velux windows were installed to light what could have been a dark access corridor.

The revised plans were finally passed by Beverley planning department in 2000 and work began in January 2001. The building was bare and basic. It had changed little since it was built in the late 1800s and had exposed brickwork, oak beams and, at one time, a slate roof – although the slates were stolen just two days after the Smarts bought the complex in the mid 1980s, and replaced with pantiles which were still in good condition. Although measuring only 15' wide, the barn is long and quite low – characteristics Arthur wanted to emphasise without losing a sense of balance.

The first task was to brick up the original large openings which faced into a stockyard, then create new openings for large, floor-to-ceiling Argon-filled windows on the opposite side of the building, overlooking the garden – also designed by Arthur. "Effectively, we turned the building round to face the other way," says Arthur. "It made sense to create the garden myself as the design of the house was tied in closely with external walls and path shapes. Originally we wanted an aluminium glazing system to remove the brick pillars between the windows – so it was effectively a wall of glass – but the planners felt it was a bridge too far and we had to compromise."

At around the same time, the brick and block extension was also built to house a studio/workshop for Tony, a sculptor. The entire roof of the barn was then removed and made structurally sound before being replaced and fitted with 400mm of insulation. A small extension was also

"THE CLEAN TACTILE LOOK CAME FROM ARCHITECT JOHN PAWSON'S BARN CONVERSIONS."

built off the main living area to house a compact library. Five Velux windows were installed – four of them in the roof over the main access corridor – to bounce as much light as possible into the house and off the plain white walls.

Arthur designed the internal block walls to accommodate wardrobe space between the two bedrooms and created two bathrooms back-to-back to keep plumbing as streamlined as possible. Rather than have radiators, which would have restricted valuable wall space, the Smarts opted for underfloor heating supplemented by electric kickboard heaters. Clean lines continue with the use of streamlined cupboards, concealed so that they look like part of the walls, and plastered brick pillars strengthen the lines of symmetry. To prevent the house looking too long and thin, Arthur created the illusion of width through a waist-high galley kitchen which hides the utility and storage areas but maintains the room's open plan feel.

"Everything we have done in this building has been simple," says Arthur, whose inspiration for the clean, tactile look came from architect John Pawson's barn conversions. Even the furniture and bathroom fitments were chosen for their crisp lines and neutral colours, along with the woodburning stove – which emits enough heat to warm the whole house.

The Smarts were so keen to maintain the clutter-free look that they didn't want to see any sockets or wires around the house, so Arthur hid all the systems within the wall cavities and switch panels inside cupboards. There are dozens of storage areas ingeniously hidden within the walls and the master bedroom hides a walk-in dressing room and small sauna – all efficiently housed behind flush-fitting closed doors.

"I have been brought up in a very artistic, creative household so I have always loved making things, planning and designing," says Arthur. "To be given the chance to design my parents' house was incredibly exciting because there was a lot of trust and understanding between us. This is a home made by the people who live in it, so it is perfectly suited for their needs. In this case, the building, their home, is a showcase for their art. You couldn't have one without the other because they fit together perfectly. And if the building does nothing more than give them a better life, then we've achieved everything we set out to do." ■

FACT FILE

Names: Arthur Smart for Anne & Tony Smart
Professions: Trainee Architect, Artist & Sculptor,
Area: East Yorkshire
House type: Converted cow barn
House size: 180m^2
Build route: Self-managed
Construction: Brick and block
Finance: Private
Build time: Jan '01 – Nov '01
Land cost: already owned (est. £100,000)
Build cost: £150,000
Total cost: £250,000
House value: £300,000+
Cost/m^2: £833

17%
COST SAVING

FLOORPLAN

The main sitting room has a vaulted ceiling making use of the barn's original oak roof trusses. Concealed cupboards in the corridor area, linking the bedrooms and workshop, provide ample storage.

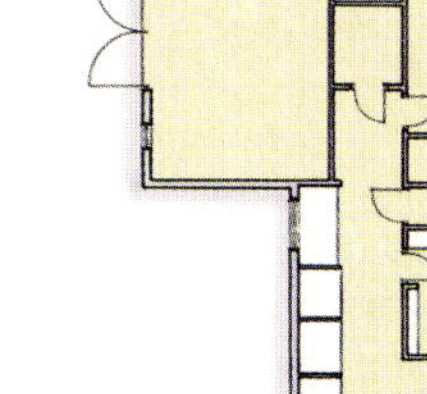

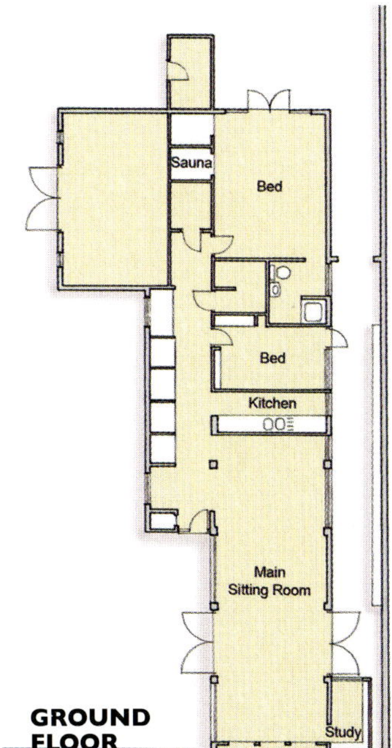

USEFUL CONTACTS

Designer – Arthur Smart, Halsall Lloyd Partnership: 0151 708 8944 or arthursmart@hotmail.com; **Anne Smart residential art courses:** 01377 217438; **Furniture and Sanitaryware:** Kris Verity at 100 per cent Contemporary: 01482 888100; **Hase Kaminofenbau (Modena) Fire:** Old Flames, **Beverley Sauna:** Ab Lagerholm: 01579 384836; **Heating System** – Firebrand Boiler (Olympic Combi): 01752 691177

A BARN REBORN

Cat and Stan Donnelly have converted a near derelict barn on a DIY basis over six years for the princely sum of just £31,000..

WORDS: CLIVE FEWINS PHOTOGRAPHY: NIGEL RIGDEN

CAT AND STAN Donnelly's barn conversion clings to the side of a narrow Herefordshire valley. Surveying the cosy, beamed, seven metre wide kitchen and its rugged inglenook fireplace, which they constructed themselves, it is hard to believe that in 1994, when they bought the property, it was home to a herd of goats. The main body of the barn was used for storage, with an upper floor that had been used to house grain. Converting the old brick and stone barn was a six year task – only because Cat and Stan chose to do 90 per cent of the work themselves and both had full-time jobs throughout the entire period.

Once they had decided to move to the countryside from their tiny terrace house in Cardiff, where Cat works as a sports ➤

Before the conversion

scientist at the university, it took them only a week to find the barn, which is one hour's drive from the city. "It had been on the market for three months and after a few days negotiating we bought it for £50,000," says Stan.

Apart from its superb position, what particularly drew Stan – a joiner by trade – to the barn was the way it had been divided up. The giant steel beams that had been inserted to support the upper floor were supported by three steel pillars. Stan realised that although it would be virtually impossible to move the pillars, all three stood in positions that could easily be worked into a downstairs layout and then disguised within walls. "I could tell both the walls and the roof structure were sound, and the presence of the steel frame inside meant we would not have to build an inner downstairs skin of blockwork to support the upper floor," explains Stan.

In return for an especially generous Christmas present, Stan persuaded his brother-in-law, a local authority draughtsman, to draw up some possible layouts, with a brief to leave large areas of the thick rough sandstone walls visible wherever possible. Where the walls are plastered Cat and Stan drylined and skimmed onto a framework of 50mm x 100mm reclaimed timber, leaving a 50mm cavity between the solid stone walls and the mineral wool insulation they inserted between the timber studs.

The kitchen is reached by two steps down from the hallway. Although it is only one storey, Cat and Stan believe it to be contemporary with the rest of the mid 18th century building. "When we came here it had a roof of sheet steel. Removing it and replacing it with reclaimed Welsh slate was one of the biggest jobs," Stan says.

Cat and Stan managed to salvage a lot of materials, such as the Belfast sink and kitchen drawers which came from a part of the local university which was being demolished. ➤

BREAD

"We decided we would rather make a feature of the hall than have a fourth bedroom," says Stan.

Creating the large full-height galleried hallway that rises right up to the roof in the centre of the three-bay barn was also a massive task. "At first we were going to have a much smaller staircase and a bedroom where there is void upstairs, but there are only two of us and we decided we would rather make a feature of the hall than have a fourth bedroom," says Cat.

By creating a full height hall in this way, Cat and Stan were able to reveal and make a feature of a large quantity of internal stone wall and to create high clerestory windows on either side of the landing at wallplate level. These draw in valuable light and offer splendid views both up and down the valley. Stan made the frames and was able to fit them comparatively easily because they are inserted into the timber framed and weatherboarded space above, where the tall barn doors once stood on either side. He and Cat were also fortunate in being able to create holes for further windows because the building is not listed. The finishing touch of this very impressive hall space is provided by the exposed massive roof timbers.

"WE INTENDED TO DO NEARLY ALL THE WORK OURSELVES... BUT WE BOTH PLANNED TO CARRY ON WORKING THROUGHOUT."

At times there were large gaps during the six year conversion when they were unable to do much work because Stan had to spend some long periods overseas where he was working as a steel fabricator. "From the outset we intended to do nearly all the work ourselves and we knew it would be a long haul as we both planned to carry on working throughout," says Cat. "Nevertheless, we were hopelessly naïve at first. It seemed to take us quite a while to realise that the barn came with no electricity, water or sewage arrangements."

Fortunately the first summer they spent working on site was a good one — they spent the first few weeks in a tent before they bought a caravan for £180. "We had no bath or shower for two years," says Cat. "Our drinking water came from our neighbours and for the first few months we went down the hill to the River Wye for a wash, then spent the evenings in the pub as we had no electricity at home."

As well as obtaining most of their timber secondhand – all the pine for the floors came from a reclaimed timber specialist in the Midlands – Cat and Stan obtained most of their washbasins and doors from reclamation yards. Also, the university was demolishing one of its science blocks, which also provided a useful source of old materials.

"We found some remarkable bargains," says Cat. "Some of our best finds, such as the art deco wrought iron bedhead which we found in a hedge and restored and now use in our spare room, were free. We had very little capital, and as we could not get a large mortgage – the state of the barn did not warrant it – we had to do everything on a shoestring. And we have succeeded!" ■

FACT FILE

Costs as of Oct 2003
Names: Cat and Stan Donnelly
Professions: Joiner/steel fabricator and sports scientist
Area: Herefordshire
House type: Three bedroomed barn conversion
House size: 204m²
Build route: Self managed
Construction: Brick and stone
Finance: £60,000 loan from Barclays
Build time: 6 years. Completed '03
Land cost: £50,000
Build cost: £31,500
Total cost: £84,500
House value: £300,000
Cost/m²: £154

72%
COST SAVING

Cost Breakdown:

Legal fees plus easement	£1,500
Water and sewage installation	£3,000
Electricity supply	£700
Welsh slates for kitchen roof	£700
Other roofing materials	£300
Floor tiles	£500
Other flooring	£600
Plumbing and gas central heating installation	£3,000
Plastering	£1,200
Electrics	£1,000
Joinery	£2,000
Insulation	£1,300
Plaster and plasterboard	£1,000
Hire fees and purchase of caravan	£500
Other materials	£5,200
Miscellaneous	£12,000
– VAT return on materials	£3,000
TOTAL	**£31,500**

FLOORPLAN

Space for a fourth bedroom at first floor level was sacrificed for a void for the double height hallway. The kitchen is situated in a single storey section of the building.

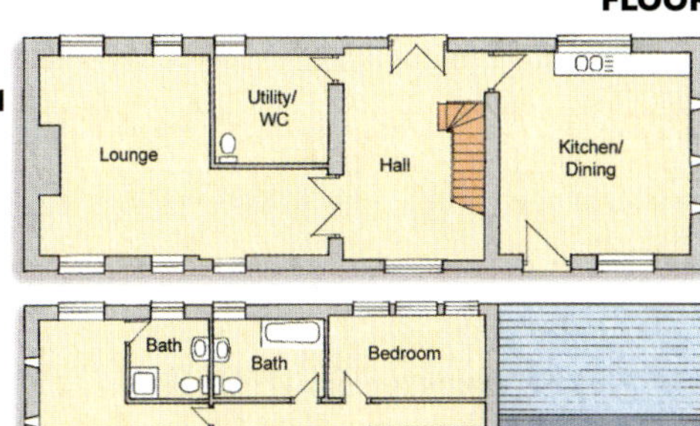

USEFUL CONTACTS

Reclaimed materials - Cardiff Reclamation: 029 20 458995; B Ravenscroft Reclamation: 01989 770269; **Reclaimed timber** - Malro: 01384 413540; **Heating and plumbing** - Simon Lewis: 01432 761434; **Joinery** - Advanced Joinery: 01981 241071

CONVERTING BARNS USING THE LOCAL VERNACULAR

LOCAL FLAVOUR

Anthony and Jenny Hudson have converted two redundant brick barns to create elegant new accommodation that draws on its Norfolk surroundings.

WORDS: CLIVE FEWINS PHOTOGRAPHY: ROB JUDGES

AWARD WINNING ARCHITECT Anthony Hudson had seen – and disliked – many barn conversions in the course of his career. So when he decided to become a self-builder and purchased two redundant brick farm buildings at the rear of his Queen Anne farmhouse in Norfolk, he was determined to do something 'different'.

If you are familiar with Anthony's work, which is firmly in the Contemporary school – his most famous commission was Baggy House in Devon in the mid 1990s – you will not generally associate him with traditional farm buildings. However, like all architects, he has always had a deep fascination with textures and, as he is Norfolk born and bred, the style of the traditional buildings of that county is firmly ingrained in his imagination.

He therefore saw the conversion of the barns into overspill accommodation for family visitors and friends and also for holiday letting as a 'fun' project that would draw him back to his Norfolk roots.

"I have seen so many conversions remove the spirit of these lovely old buildings," he says. "So many of them end up completely losing their original charm and character. Generally this is because people will insist on destroying much of the original and covering up the rest."

Anthony has deliberately set about retaining what he calls the "strong forms" of the barns by only building one very shallow extension, inserting very few small new windows and making the internal layout fit neatly into the original structure. He has retained virtually all the original ➤

The unusual floor is called 'Perstorp' and consists of marble chips laid in resin, creating a chalk-like appearance.

Single storey Quaker Barn, the smaller of the two buildings, is used as a self-contained holiday let. The roof space has been left open to retain the building's original character.

"AS FAR AS POSSIBLE I WANTED TO GO WITH THE IMPERFECTIONS IN THE OLD BUILDINGS. TRADITIONAL NORFOLK BUILDINGS ARE FULL OF A PATCHWORK OF WONDERFUL COLOURS AND TEXTURES."

timbers and made sure as many as possible remain visible.

"My wife Jenny and I wanted to keep the feel of agricultural buildings and not domesticate Hall Barn or its neighbour Quaker Barn too much," Anthony says. "I did not even wish to use any modern rooflight systems, although I am sure this would have been permitted." Instead he used groups of overlapping glass pantiles – all reclaimed, as he could not find a manufacturer of new ones – with double glazed units beneath. At the same time he wanted to create interiors that were light and airy – and also to produce something that as well as being what he calls "a bit of fun," would satisfy him professionally.

Anthony could see that Hall Barn, the larger of the two, which had served as a cowshed and was open to the front, had the height to accommodate two upstairs bedrooms. There are two more bedrooms downstairs.

To avoid, as he calls it, "peppering" the building with too many windows, he devised a new form of 'frameless' window system that makes the most of existing small openings. It is fitted to the outside of the structure and comprises double glazed sliding units in simple steel tracks, draughtproofed with car window seals. "This system both keeps the openings small and looks industrial. I feel it suits the look of the barn rather than timber windows, which we have kept to the new section at the rear," Anthony says.

To try to keep costs down Jenny acted as client in order to restrain Anthony where necessary. The Hudsons saved money by using reclaimed materials as much as possible and sourcing most of the new materials locally. The straw bales, oak joinery and flint were all sourced within a five mile radius of the site and the man who acted as site manager, Brian Buck, and most of his team all live locally. Both of the sturdy black log burning stoves came from a local reclamation specialist, while the solid oak front doors are made from oak boards sourced locally, fixed to solid ply planks.

"As far as possible I wanted to go with the imperfections in the old buildings. Traditional Norfolk buildings are full of a patchwork of wonderful ➤

The 25m long straw bale wall in the larger Hall Barn is thick enough to allow deep window reveals. It has remarkable insulation qualities and cost a total of just £1,000.

colours and textures and we have tried to add to these rather than remove them," Anthony says.

In the smaller Quaker Barn, which serves as a holiday let, he has used local flint for an internal wall that divides the kitchen and dining area from the bedrooms. The best example of his use of local materials, however, is the 25 metre long, five metre high solid straw bale wall on the north side of Hall Barn. Here Anthony has retained all the vertical oak posts – the building was a hay store – and filled the spaces between with the straw bales, which are fixed to steel frames that span between the oak posts. The outside is shielded by a rain screen of translucent glass fibre panels.

At eaves level the straw bales, which are laid in English bond pattern and spiked together with 10mm steel bars, are screened with horizontal green oak boards to create the effect of the original cladding of the building at the front.

"It is 600mm thick and the whole structure, including the oak boarding above, the steel framing and the stainless steel gauze placed at the bottom to prevent rats from entering the straw, cost less than £1,000," Anthony explains. The glass fibre panels for the rainscreen cost £400.

"You can achieve tremendous effects very cheaply with straw bales, but you have to have the space – they are very thick – and use them in the right place," Anthony says. "Straw bale walls are very ecological because they are recyclable. They are also very cheap and they have tremendous thermal mass – the effect is one of permanent warmth when you place your hand against the wall."

The great depth of the walls has provided deep window reveals which are glazed with sandblasted glass. The windows, which are not immediately apparent from the outside, are all strictly rectangular and different in size – and are placed at different heights to create an interesting play of light. They are glazed internally with opaque glass. On the inside of the bales there is a 50mm lime plaster which has been left in its natural colour.

On the south side Anthony has thrust the building out by a metre by adding

"YOU CAN ACHIEVE TREMENDOUS EFFECTS VERY CHEAPLY WITH STRAW BALES, BUT YOU HAVE TO HAVE THE SPACE – THEY ARE VERY THICK – AND USE THEM IN THE RIGHT PLACE."

a shallow rectangular bay window in rendered blockwork with sliding oak windows and a weatherboarded external wall.

This imports a lot of light and makes the most of the views over rolling fields and copses. It has a flat roof and is a play on Anthony's Modernist tendencies, complete with a steel chimney containing the flue from the log burner, although the rest of the heating is under the floor.

The bay is the most radical change in the building. To create it the builders had to remove one of the posts and a tie beam from the six bay building but otherwise the frame is undisturbed. The oak sliding windows here, made by a local company, were the only really expensive component apart from the sanitaryware.

Other ingenious touches on the outside include the brick wall outside the downstairs WC and shower room. Here no new window has been created but in order to import light into the bathroom a traditional barn 'fretwork window' in brick has been formed into the outside wall. This is a traditional external feature in Norfolk brick barns but conceals a window behind that imports light into the bathroom.

Inside, the dominant feature is the long steel bridge that crosses the central relaxation area, straddling one of the tie oak beams in the process and linking the two upstairs bedrooms. It was made by a local steel fabricator, as were the stair balustrades.

Although the inside has a crisp Modernist feel, Anthony has introduced a 'rough' floor made from marble chips laid in resin. "It is called Perstorp," says Anthony. "It suggests chalk, which is very prevalent in this part of Norfolk – again I had local connections in mind."

The adjoining Quaker Barn is a one storey building, but because it is built on a slope, Anthony has managed to build in a step up to the kitchen and master bedroom. Light again pours in from the south through a quadrant in the end gable and a large window in the wall beneath.

"I aimed to complete the dual project using local resources and labour, sustainable materials and adding very few modern insertions," Anthony says. The policy seems to have paid off, because the project recently won a regional RIBA Award. ■

FACT FILE

Names: Anthony and Jenny Hudson
Profession: Architect + potter
Area: Norfolk
House type: Four bedroomed two storey brick barn
House size: (Hall Barn) 175m²
Construction: Brick, with straw bale wall on site of previous open front
Build route: Self-managed
Finance: Private + Barclays
Build time: Jun '00 – Aug '01
Barn cost: £30,000
Build cost: £130,000
Total cost: £160,000
House value: £350,000
Cost/m²: £743
(figures relate to Hall Barn)

55%
COST SAVING

GROUND FLOOR

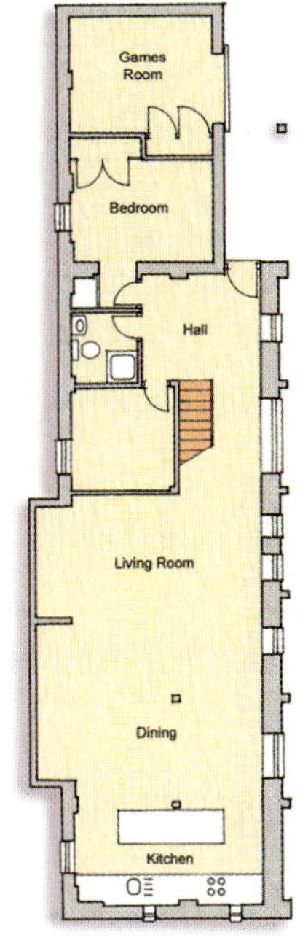

FIRST FLOOR

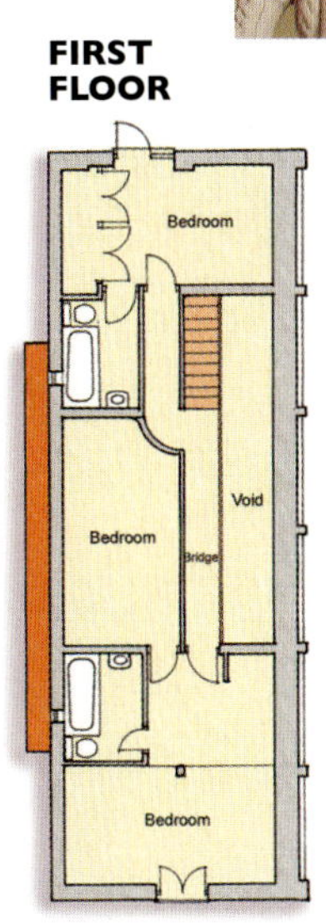

FLOORPLAN

The open plan living/dining/kitchen area is overlooked by a bridge, with three bedrooms (one en suite) to the first floor.

USEFUL CONTACTS

Architect – Hudson Architects: 020 7490 3411; **Structural engineers** – Alcock Lees partnership: 01603 764448; **Site manager** – Brian Buck: 01362 688531; **Glass fibre cladding** – Filon 0121 353 0814; **Downstairs floors** – Laser Contracts: 01359 271417; **Electrics** – Ray Riseborough: 01603 871437; **Underfloor heating** – Richard Brown: 01603 898904; **Reclaimed pantiles** – Norfolk Stoves and Fireplaces: 01603 860762; **Metal fabricators** – Stanley Welding Services: 01603 873099; NB: For details of holiday lets in the Hudsons' barns contact Rural Retreats: 01386 701177 www.ruralretreats.co.uk

CONVERTING A STONE STABLE AND BARN INTO A NEW HOME

TWO BECAME ONE

Alex and Annemarie Hole have converted a barn into a stylish new home which mixes their needs with its existing character.

WORDS: JASON ORME   PHOTOGRAPHY: JEREMY PHILLIPS

IT'S RATHER EASY to get lost in the rural network of narrow lanes, small hamlets and large, open expanses of farmland to the south of Newark, Nottinghamshire. H&R found the journey to Alex and Annemarie Hole's barn conversion totally disorientating – but the same mystifying turns proved much more fortuitous for them.

"I took a wrong turn on the way to work one day and came across a man putting a 'For Sale' board outside a couple of barns," explains Alex. "I got out of the car to take a look around and realised that, with the combination of the serene views and the position, there was a lot of potential."

Alex took his partner, Annemarie, out to have a look that night and within a couple of weeks they had bought the place. "At the time, the two barns would have been too much for us to take on individually, so we encouraged some friends to take a look around and they fell in love with the place, too. It was brilliant having our friends so close during the project and while we kept all of the ordering of materials totally separate, we were able to swap labour when we needed to."

Alex and Annemarie run a successful development company which they formed after undertaking a project for their first home together – the conversion of some rather unfashionable 1960s office space into an attractive flat in Nottingham city centre. The company has grown and developed a reputation and expertise in converting barns – which is a good thing, because the area has lots of barns and hardly any building plots.

"I'd developed a good connection with the planners on my previous development projects," says Alex, "and they fully supported our scheme. We wanted to keep the south facing/road side of the property exactly the same, while on the private north facing elevations, which looks out over open countryside, we've included more openings. The south facing doors are in fact large expanses of glass anyway – I did

"SOME PEOPLE... END UP WITH SOMETHING THAT IS LESS OF A HOME THAN A MUSEUM..."

The south facing side of the kitchen (left in this picture) contains purely the original openings, while on the private north facing side Alex and Annemarie were permitted to introduce new windows.

Alex's office, a new construction in the space filling the corner between the original barn and the stable, contains lots of glass to take in the views over the countryside.

that because I wanted them to appear as if they were open from the road."

Converting barns is an exercise in compromise – of balancing out the different weighing concerns in a manner that new-builders don't have to deal with. "So many people fail because they try and force their own design ideas on the barn so that it becomes, from the inside at least, like a standard house, full of straight lines and no space; equally, some people are so scared of disturbing the original character that they actually end up with something that is less of a home than a museum. We wanted to pick the best bits of each approach," he explains.

The stable part of the building was linked to the threshing barn by a very narrow walkway and Alex knew that, to make the sections work as a whole, he would have to undertake some additional building to fill in the corner. The result is a glazed office space that is much more contemporary in feel than the rest of the building but, as it is hidden from

"PEOPLE THINK THAT CONVERSIONS ARE CHEAPER THAN NEW BUILDS SIMPLY BECAUSE THERE ARE SOME EXTERNAL WALLS ALREADY IN PLACE." ➤

The living room, in the threshing barn section of the home, is complemented by the addition of the new structural posts and beams which support the first floor.

The en suite bathroom has underfloor heating as the rest of the home has fitted carpet.

"WE STARTED IN JANUARY 2002 AND MADE THE BIG MISTAKE OF PROMISING TO OUR FAMILY THAT WE'D GIVE THEM A BIG CHRISTMAS DINNER IN OUR NEW HOME."

public view, enables Alex to enjoy the views without necessarily affecting the rural character of the barn.

Originally a surveyor, Alex knew that the existing walls of the threshing barn section of the L-shaped building would be forced to bow out by the strains placed upon them by a first floor and so designed a series of pillars and braces which are exposed in the large living area.

"One of the major debates for us was between an open plan and more cosy spaces," says Annemarie. "I didn't want to live in a couple of large open voids but at the same time I didn't want the rooms to be so small that they could have been in a conventional new built home."

The compromise seems to have been as much in the specification of the fittings as in the design. Not many self-builders with an eye for design are brave enough to admit that they find wooden flooring in the living and bedroom spaces too cold and, as they like nothing better than nice warm feet and the occasional stretch out on the floor, specify carpet. It also makes the larger spaces feel – that word again – cosier, and it is easy to be persuaded by their argument. With heavy flagstones in the kitchen and slate in the utility, Alex and Annemarie seem to have hit on a simple design point – flooring should be mixed to meet the different needs of the individual room.

"We considered underfloor heating but as I have never used it before, and with the choice of carpets, I decided to play safe," admits Alex. "We have installed an electric underfloor system in the en suite, however,

which is simply wonderful. It was cheap to install, is highly controllable and, most importantly, keeps our feet warm."

The project took a little over a year to complete. "We started in January 2002 and made the big mistake of promising to our family that we'd give them a big Christmas dinner in our new home," laughs Annemarie. "Despite being in, however, we were far from finished!"

"People think that conversions are cheaper than new builds simply because there are some external walls already in place," says Alex. "That couldn't be further from the truth. Those walls actually become a problem rather than a benefit and the time taken to deal with the nuances any old building throws up – from adding insulation and damp-proofing to an existing floor, to the non-standard window sizes requiring everything to be made to order – mean that they are invariably more expensive. The offset, of course, is that you end up with a home of instant character in the most scenic of isolated locations." ■

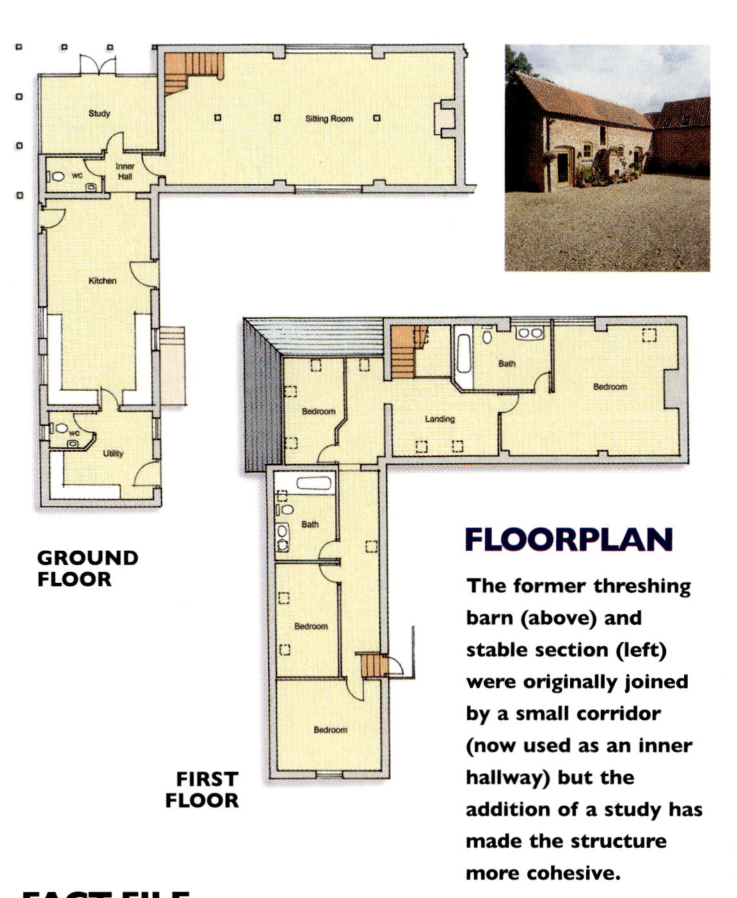

FLOORPLAN

The former threshing barn (above) and stable section (left) were originally joined by a small corridor (now used as an inner hallway) but the addition of a study has made the structure more cohesive.

FACT FILE

Names: Alex and Annemarie Hole
Professions: Developer and scientist
Area: Nottinghamshire
House type: Barn conversion
House size: 320m²
Build route: Self-managed
Warranty: Building Surveyor's Certificate
Finance: Private
Build time: Jan '02 – April '03
Property cost: £200,000
Build cost: £220,000
Total cost: £420,000
House value: £600,000
Cost/m²: £687

37%
COST SAVING

USEFUL CONTACTS

Design and build – Willow Tree Developments: 01949 851510; **Building surveyor** – Robert Walker Associates: robert@robertwalkerassociates.fsbusiness.co.uk; **Purpose made joinery** – FA North: 0115 987 2339; **Kitchen units** – Chris Ablewhite Joinery: 01949 861679; **Bathrooms** – Bathhouse Design: 01522 869088; **Tiles and slate flooring** – Studio Ceramics: 01636 673527; **Plumbing** – Carpmail Heating: 0115 939 8719; **Electrics** – Premier Electrical Services Ltd: 01509 856956; **Undertile heating** – Warmup: 0845 345 2288; **Insulation** – Kingspan: 0870 850 8555

The roof, which dominates the front elevation, combines rustic clay tiles and varying pitches with large glazed sections, for a modern interpretation on traditional early 20th century design.

NATURAL BEAUTY

David and Sophie Harber have built an oak framed home with the main living spaces on the first floor.

WORDS: CLIVE FEWINS PHOTOGRAPHY: BRUCE HARBER

The kitchen, along with the other main living spaces, is situated on the first floor.

DAVID HARBER WAS very keen to have an oak front door as the finishing touch to his new 'upside down' oak framed house in a pretty village in South Oxfordshire. The only problem was that he could not afford it. He and his wife Sophie were close to overrunning their original £120,000 budget by some £100,000 – so he decided to make it himself.

With the aid of an old friend who was staying with them, he decided to do it on the day of their housewarming party. He had already made the steel frame at his workshop near Henley-on-Thames, where he runs a business making sundials, and the plan was to fix horizontal oak planks inside and out.

"Despite having set ourselves an impossible schedule of about half a day we managed to fix the oak planks and hang the door just as the first guests were turning into our drive," says David. "It was a useful deadline because the house had been rather insecure as it was – with a piece of mdf as a front door since we moved in a few weeks earlier!"

Ever since they married seven years ago, David and Sophie dreamed of building a contemporary oak-framed house with all the living rooms upstairs. "We had seen and loved a number of houses like this and liked their informality, which we always felt suited our lifestyle," Sophie says. "We also thought it more fun to live upstairs as all the interesting timberwork is in the roof and you get better views."

Where the roof sweeps over the hall at a steep pitch it is entirely glazed, a feature which continues outside to form a porch — one of several elements in the house that still thrills the Harbers.

Glazed section of roof let in light and give the illusion of extra height

"WE ALSO THOUGHT IT MORE FUN TO LIVE UPSTAIRS AS ALL THE INTERESTING TIMBERWORK IS IN THE ROOF AND YOU GET BETTER VIEWS."

As soon as you spot the house, tucked away in a quiet tree-lined position in the heart of the village on a site which formerly housed a small bungalow, it is obvious that the roof is a tour de force. It is large, with rugged clay tiles and a variety of pitches, and sweeps down to eaves height in several places in a fashion that imitates the classic English style of the early 20th century. It is given a modern interpretation by several large glazed sections that make the interior a mass of light.

There are many other imaginative features, not least the wealth of sundials on show in the garden and on the outside walls. David and Sophie were also very instrumental in the design of the balcony, which is cut into the roof like a reverse dormer and has glass cheeks that separate it from the main living area. "The roof sweeps over it so that you can sit out and read or eat in the rain on a mild day, even in autumn or early spring," says David.

The garden faces west and enjoys the evening sun, while much of the house catches shafts of sunlight during the day because of the roof glazing and the large areas of glass in the walls.

In many ways, however, the piéce de resistance of the house is the external staircase of green oak that leads straight down to the garden. "Living upstairs you can easily feel you lose touch with the garden and that was the last thing we wanted," says David, "especially as it plays such an important role in our lives, as it is an outdoor showroom for the sundial business that supports us."

"Although it catches much of the sun on this garden-facing side, we ➤

would have preferred the whole house to face south," says Sophie. "This would have given us a more secluded outlook but the Vale of White Horse planners would not allow it. However we were allowed to increase the size of the original tiny bungalow from $83m^2$ to $240m^2$, which we felt was quite generous."

At first the Harbers thought of expanding and rebuilding the existing property. Then, when they realised that the project would be zero-rated for the purposes of VAT if they demolished to foundation level, they decided to take this course and try to build as far as possible to the original foundations, adding only concrete pads where needed to support the main full-height oak posts.

"It was one of the last plots in the heart of the village, which is one of enormous character and a place we were very keen to live in," David says. "We had been looking for ages, so wanted to make the most of the opportunity."

To achieve this quality of build and also such a large and expensive roof the Harbers had to handle their budget very carefully. Their first plan had been to get the house built and live with a temporary kitchen and minimal decoration until they had enough money to finish the job.

"Then the bank allowed us to extend the mortgage and we thought it

"WE LOVE SOME OF THE MORE INTERESTING INTERNAL FEATURES LIKE THE FIREPLACE HE DESIGNED AND THE 'SECRET TUNNEL' THROUGH THE WALL AT THE BACK OF THE GIRLS' WARDROBES."

would be far fairer to our three small children to finish it off as soon as possible," Sophie says. One way in which they managed to save money was by David building the main staircase. Like the external stair it has solid oak treads, which David had made at a local sawmill. He bolted them to steel supports and made the balustrading out of surgical grade stainless steel angled sections, which he cut into an irregular outline using a plasma welder. He then cleaned it up and installed tensioned stainless steel wire as balustrading.

Now it is all finished there are several things the Harbers would like to change. "I'd really have liked a fifth bedroom, but we could not afford one at the time," Sophie says. "There is also the problem of the hallway downstairs. It is beneath the upstairs landing and rather dark because the small window at the far end is overshadowed by the balcony. David and I had an argument about this section of the downstairs and he won. Really it was part of our economies – basing the design on the existing foundations – and was originally meant to be a playroom for the children. We have found it is far more convenient to use the upstairs room adjacent to the kitchen for this purpose rather than the office it was designed for, so we may eventually just reverse roles and turn this section into a home office by placing a glass screen between it and the rest of the hall."

"Overall we are delighted with the house. We were incredibly lucky to find our architect – Gary Bain, who was extremely amenable – and we love some of the more interesting internal features like the fireplace he designed and the 'secret tunnel' through the wall at the back of the girls' wardrobes. We could have built very much cheaper if we had decided on a simpler single pitch roof, but to get what we really wanted demanded a complicated roof and we think the result has been well worth it." ■

FACT FILE

Name: David and Sophie Harber
Profession: Directors of sundial manufacturers
Area: Oxon
House type: Four bed detached
House size: 240m²
Build route: Main contractor + selves
Construction: Oak frame with rendered blockwork on ground floor
Warranty: Architect's Certificate
Sap rating: 87
Finance: Private + £300,000 Lloyds TSB
Build time: Nine months

Land cost: £270,000
Build cost: £250,000
Total cost: £520,000
House value: £800,000
Cost/m²: £1,041

35%
COST SAVING

Cost Breakdown:

Timber frame	£52,000
Main contractor	£132,000
Fees and connections	£12,000
Roof and tiling	£9,400
Plumbing and heating	£9,000
Electrics	£8,600
Bathrooms	£4,800
Kitchen	£5,500
Glazing	£15,000
Misc	£1,700
TOTAL	**£250,000**

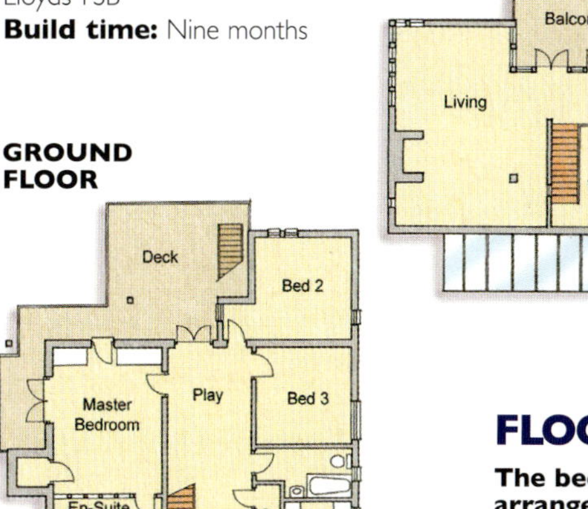

FLOORPLAN

The bedrooms are arranged on the ground floor while the main living area is on the first floor. The roof has several large glazed sections to let the light in.

USEFUL CONTACTS

David Harber Sundials: 01491 576956; www.davidharbersundials.com; **Architect** – Gary Bain: 01962 715021; **Timber frame and erection** – The Timber Frame Company: 01458 224463; **Kitchen** – John Lewis of Hungerford: 01924 380038; **Carpentry and plastering contractors** – McLellan Builders: 01389 830231; **Plumber** – Rick Taplin: 01491 837982; **Electrics** – Peter Witney: 01491 837939 **Bathroom** – The Limestone Gallery: 01488 688103; **Rooftiles** –Tudor Rooftiles: 01797 320202

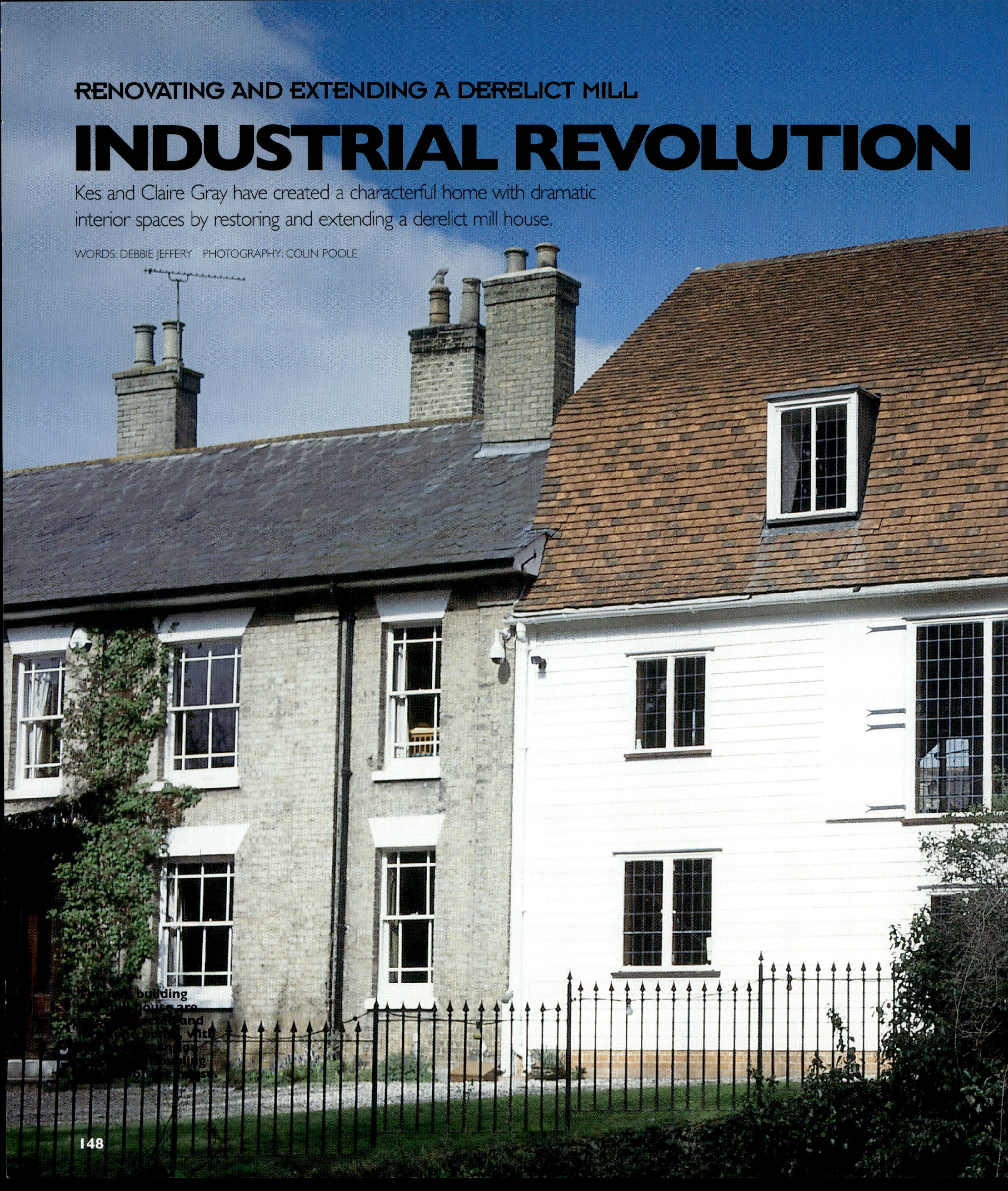

RENOVATING AND EXTENDING A DERELICT MILL

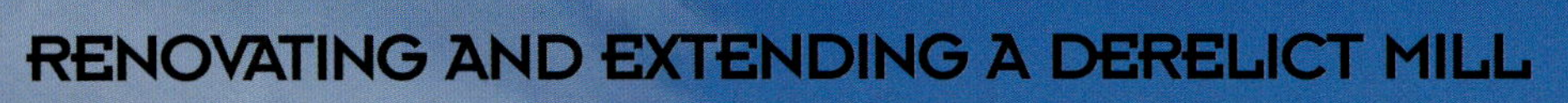

INDUSTRIAL REVOLUTION

Kes and Claire Gray have created a characterful home with dramatic interior spaces by restoring and extending a derelict mill house.

WORDS: DEBBIE JEFFERY PHOTOGRAPHY: COLIN POOLE

The mill building and mill house are [illegible] and [illegible] one home, with [illegible] extension [illegible] doubling [illegible] of the house.

MY FATHER BEGGED me not to buy The Mill House," says Kes Gray. "It had been empty for a number of years and was in a terrible state. Vandals had stripped it and sprayed graffiti, kicking through the lathe and plaster ceilings with their feet. They had used floorboards to smash the roof slates and the remaining shell was frequented by the local glue sniffers, who had filled the mill pond with their plastic bags. There was excrement all over the walls – it certainly wasn't anyone's idea of a dream house!"

Photographs confirm the appalling destruction which faced the Kes and Claire when they decided to view the property. Boarded up windows stare blankly from its ivy covered façade and the interiors are strewn with debris and look distinctly unsafe. Kes could see past the degradation however, and decided to buy the three bedroom brick house. "From a young age I had always wanted to live in a mill," he explains, "and a structural survey revealed that the building itself was sound – it just needed some TLC. I decided to ignore my father's advice!"

The Grays, who met at school, had both lived in Chelmsford as children and knew the area well. After marrying they had moved to London, where they lived for eight years, but decided to return to Essex with their children in order to be closer to their family. Kes telephoned estate agents requesting information on houses requiring renovation, and was sent details of the 1856 mill house at Witham. "We didn't need a key to get in," says Claire, "and there was a sign saying 'this house is haunted'. Nobody could believe that we wanted to buy it."

Builders had purchased the house and had been trying for some time to gain planning permission to build on the land. By allowing the property to fall into dereliction they

The massive sitting room fills the entire dowstairs of the extension which has a footprint similar to that of the original mill – and a ceiling height of 4.75m. Upstairs there is further floor.

"BUILDING THE MILL WAS A DREAM WHICH WE HARBOURED FOR SEVERAL YEARS WHILST RENOVATING THE MILL HOUSE..." ➤

The unusual staircase is reminiscent of a mill wheel and leads up into the first floor of the old mill house.

hoped to be able to demolish it, but the local conservation officer was adamant that The Mill House remain. Other locals were less welcoming to the Gray family. "The children continued to try and use the building as a den, and made it clear that we were not wanted," says Kes. "We had materials stolen and discovered a hod hidden in a hollow tree in the garden which they were using to carry off our bricks. I chased some youths off the site in my wellington boots only to fall headlong into stinging nettles. It was hard work."

The Grays employed a builder to renovate the house, and cleared the mill pond of debris. The property was cleared and it took six months to restore the basic building, including fitting new sash windows, rebuilding floors, rewiring and fixing the roof. During this time Kes and Claire gathered information about the history of their new home and discovered that there had been flour mills on the site since Saxon times. "They always used to burn down, and the last mill was destroyed in a fire in 1882," Claire explains. "Kes and I found an article, written in the 1930s, which was illustrated by a photograph of the old mill."

The building had two pairs of stones worked by water and two by a six horsepower steam engine, and was occupied by a Mr Smith who, after working until 11.30 one night, banked the furnace and closed the mill, when a spark from the furnace started the fire. "In the night smoke was smelled in the house and by morning everything was gone, with only a little furniture saved" reads the article. "Two storey weatherboard, with gabled mansard roof and without a lucam, it adjoined the later mill house which survives, with traces of fire still discernible."

➤

The Grays decided that they would extend the miller's house and re-create the old mill — albeit without a working water wheel. They approached local architect Rodney Black with the blurred photograph for reference and some of their own sketches, and explained that they wanted their extension to retain the surviving wheel pit and leat. "They were very keen that the new extension should mirror the former mill in proportion, character and style – occupying its footprint," explains Rodney Black, who was recommended by the local authority. "We worked closely with the planning department, county architects and archaeologists to agree the design."

Claire and Kes researched the topic thoroughly, consulting archives, scouring literature and quizzing museums. The new building has a handmade brick plinth, a heavy Douglas fir and pitch pine timber frame, a gabled mansard roof and a large chimney built of handmade bricks. On the ground floor there is a huge room measuring 15 metres in length and 4.75 metres high which the Grays use as their sitting room, with five bedrooms and two bathrooms set into the mansard roof above.

"We wanted to make sure that the new extension would work in conjunction with the existing house and not make it redundant," Claire explains. "Keeping the kitchen and our bedroom in the older property makes us use the entire building. The extension has taken our living space into the garden and it is surrounded by trees and wildlife. It more than doubles the size of the house."

The extension has been built in timber frame and clad in weatherboarding to replicate the original building.

Located within a Conservation Area, the site is of historic interest, and an archaeological watching brief was necessary when the foundations were dug. A 17th century brick floor and the remains of a wall from the former 19th century mill were discovered, as well as burnt charcoal deposits associated with the fire. The foundation trenches straddle the mill race, wheel pit and bypass channel to the former mill (collectively known as the 'mill reach') and have been built on chalky boulder clay with patches of sand and gravel. This necessitated deep piled foundations with ground beam trenches dug between the lines of concrete piles.

"Timber frame was the only way of truly replicating the style and feel of a Georgian mill," Kes explains. "Rodney recommended our builders, who constructed the frame on site. The principal timbers were absolutely

"THE EXTENSION HAS TAKEN OUR LIVING SPACE INTO THE GARDEN AND IT IS SURROUNDED BY TREES AND WILDLIFE. IT MORE THAN DOUBLES THE SIZE OF THE HOUSE."

enormous. Planning requirements stipulated a Kent peg roof and single glazed leaded light windows, copied from the photograph for authenticity, with Czechoslovakian handmade glass which gives a gently rippled effect." High levels of insulation compensate for any heat loss, and there is underfloor heating beneath the Indian sandstone floor in the living room.

The internal walls have been finished in an unusual way, using Lux soap flakes mixed with water to create a creamy paste and trowelled onto the walls. Rodney Black pioneered this old fashioned method on historic buildings in Japan, and decided that it would work well over the mill's off-white plaster finish, creating a smooth waxy sheen once it was buffed and polished.

"Although we couldn't have a working water wheel we asked Rodney to devise a feature staircase in the extension which would be reminiscent of the original wheel, without looking too twee," says Claire. "He also designed the bold counterbalanced lighting which resembles giant mobiles and was made by a blacksmith. It suits such a large space and has a crude industrial feel."

"Building the mill was a romantic dream which we harboured for several years whilst renovating the mill house," says Kes, senior copywriter at Saatchi and Saatchi and a well-known children's author. "The mill was missing from the village for 120 years and we wanted to put it back – although we won't be making any bread for the parish! Even the blacksmith, who has lived in the village for over thirty years, said that he couldn't remember our mill not being there.

"We still have several bathrooms and bedrooms to finish and decorate but we're in no rush – we could never leave this house. People thought we were mad when we bought it as a derelict shell and it certainly made no financial sense at all – we've mortgaged ourselves to the hilt – but the children now have a four acre garden where they can play safely and we have excavated a small lake for them to fish and swim in. It really is an inspirational place to live." ■

FACT FILE

Names: Kes and Claire Gray
Professions: Children's author and chartered accountant
Area: Essex
House type: Eight bedroom renovation and new build
House size: 346m² (125m² house + 221m² extension)
Build route: Contractor
Construction: Timber frame (extension)
Warranty: Architect's Certificate
Sap rating: 97
Finance: Woolwich
Build time: July '00 - May '01
House cost: £210,000
Renovation cost: £100,000
Extension cost: £295,000
Total cost: £605,000
House value: £900,000+
Cost/m²: £1,335 (extension only)

33%
COST SAVING

FLOORPLAN

The huge double height sitting room takes up the entire ground floor of the extension, with five bedrooms and two bathrooms set into the mansard roof above.

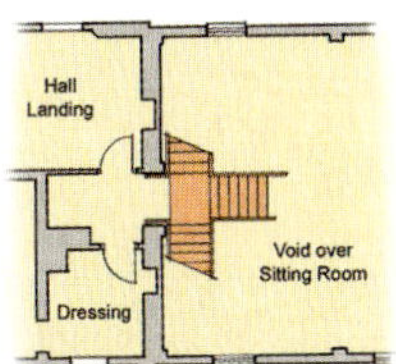

MEZZANINE/STAIR DETAIL

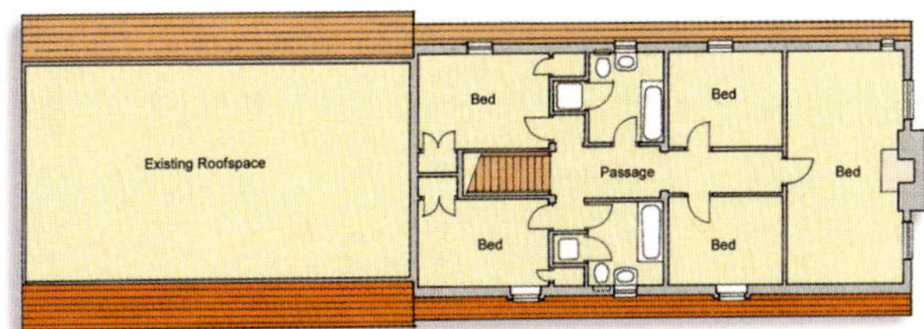

FIRST FLOOR

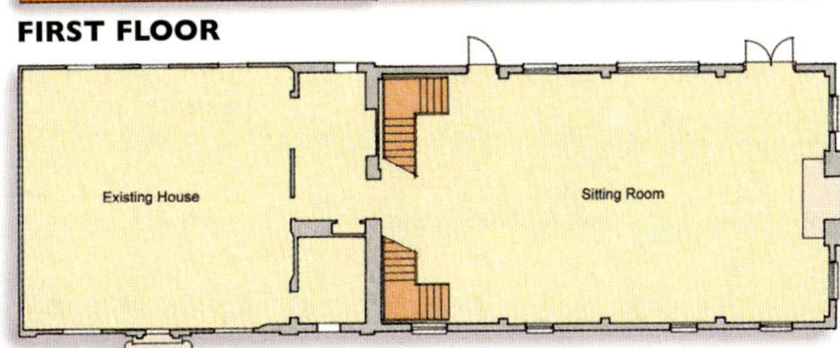

GROUND FLOOR

USEFUL CONTACTS

Architect - Geary and Black Limited: 01206 322800; **Main contractor** - Rose Builders: 01206 392613; **Structural engineer** - Ken Rush Associates: 01376 326789; **Mechanical services** - Cooper and Brome Contracts: 01206 212582; **Electrical services** - Lock Electrical: 01206 262959; **Piling** - Colets Piling Limited: 01245 224255; **Staircase arches** - Kingston Craftsmen: 01482 225171; **Roof tiles** - Tudor Roof Tile Co. Ltd.: 01797 320202

COUNTRY CHARACTER

Geoff and Sue Gore have created a new home with some subtly tasteful interiors by converting a former milking shed.

WORDS: VICTORIA JENKINS PHOTOGRAPHY: HUGH BURDEN

IN ITS ORIGINAL state, Sue and Geoff Gore's new home near Bath was a stone-built milking shed with earth floors and a piggery across the concrete yard. Despite that it had a Grade II listing (and was also in a Conservation Area), by the time the Gores saw it, a builder had begun the process of converting it into a family dwelling.

"When we went to see it, we were getting a bit desperate," says Sue. "We had sold our own home and were living in rented accommodation as the sale of the property we were buying fell through the day we moved out of our house. My first impressions of the conversion were that we just couldn't cope with that much work. The inside was dark and bleak with the breeze block partitions the builder had put in, and the outside looked like the local tip." In addition, although all the services were to be newly laid, there was to be no main drainage but a Titan septic tank instead.

"When we first saw it, the yard was an expanse of muddy concrete," says Sue. "As for the garden at the back, it was a wilderness – luckily now nearly tamed."

However, despite its drawbacks, the Gores could see the potential beauty of the building. The exterior walls were of Bath stone with terracotta roof tiles, while inside there were double height ceilings and a dramatic network of roof beams. "The cowshed is 100 years old at least and I'm assured the beams are original," says Sue. ➤

Two years on and the Gores are living in a long ranch style home, the extension built of the original Bath stone, reclaimed from one of the exterior walls which was dismantled in place of an interior wall of breeze blocks. All the breeze block walls have now been plastered over and the earth floors have vanished under smart quarried stone tiles.

"The 'new' roof tiles had to be of reclaimed terracotta to match the existing ones," Sue adds. "Likewise, the stone for the extension had to be reclaimed Bath stone, plus we couldn't put in new window openings but had to adapt those already there. All the window frames had to be handmade – even though the windows themselves are double glazed." This is because planning permission was granted on condition that only the existing openings were utilised. However, what prevents the place from feeling anything like a bungalow are the tall ceilings and the elaborate beamwork.

As the garden was too small to interest a landscape gardener, the Gores managed to persuade a friend with a digger to remove the lumps of concrete and other rubble from it. The same friend then levelled the ground, installed the decking behind the kitchen, put down 70 tons of topsoil, built the walls, laid the lawns and the gravel paths and then he and Sue planted the garden.

"What we found so great about the conversion was that we felt we could stamp our own personality on it," she says. "The kitchen was particularly important to me as I work as a chef and love to cook and entertain friends and family. It was in here and the hallway that we chose to have limestone floors for which we paid on completion of the sale. As soon as I saw the travertine floor tiles I knew they were what I was looking for. They are a naturally quarried stone, in this case from Turkey, and are light and hard-wearing – just right for those areas which have frequent use."

The kitchen units are in maple with long stainless steel handles for a cool sleek look, while the work surfaces are in black granite. "We chose two built-in eye-level ovens so we don't have to do any bending down and we have slide-out shelves in the floor units for the same reason," says Sue.

Although they sealed the travertine stone tiles behind the hob three times, they still continued to get splattered with cooking fat which was difficult to remove, so the Gores had a back panel of toughened glass from ➤

"WE WOULDN'T HAVE TAKEN ON THE CONVERSION OURSELVES... DOING IT WITH A BUILDER WORKED."

The contemporary kitchen, consisting of maple units, granite worktop and stainless steel handles, is from Simply Kitchens (01752 766866).

As the original openings in the barn could not be changed, neutral colours have made the most of the available light.

Roman Glass put in to protect them. "It is now very easy to keep clean," says Sue.

The bathroom and en suite shower room are particularly attractive as Sue and Geoff chose an engineered quartz tile from Arena Stone for the walls and floors. "It sparkles and glitters in the light and really lifts the rooms," says Sue. "We wanted a power shower in the en suite bathroom as this is what I became used to during the time I spent in Australia, and this one is like standing under a waterfall.

"As for the multi-functional room it could be a dining room, a third bedroom or an office, but we wanted it to be all three. It's a dining room when I am entertaining, a bedroom when I have extra guests and it's where the computer sits, which we visit almost every day – so we think we get pretty good value from this room.

"We wouldn't have taken on the conversion ourselves, but doing it with a builder worked well for us. We both love our new home as it has the light airy contemporary feel we wanted. With its high ceilings and wonderful beams, it retains something of its original character and we're really looking forward to entertaining friends and family this summer." ■

FACT FILE

Names: Sue and Geoff Gore
Professions: Retired banker and caterer
Area: Somerset
House type: Three bedroom barn conversion
House size: 150m²
Build route: Main contractor
Construction: Bath stone
Finance: Private
Build time: 18 months
Land cost: £200,000
Build cost: £150,000
Total cost: £350,000
House value: £450,000
Cost/m²: £1,000

22%
COST SAVING

FLOORPLAN

The extension has been utilised as a living room, with a kitchen diner and three bedrooms making up the remainder of the single storey accommodation.

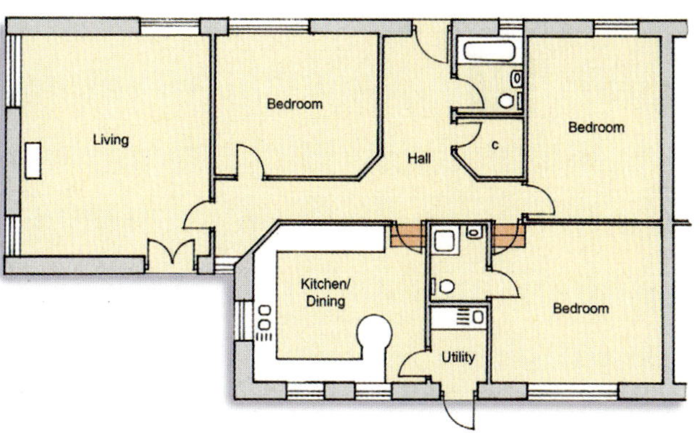

USEFUL CONTACTS

Builder – DAR Developments: 01225 316705; **Travertine floor tiles** – Mandarin: 01225 460033; **Fire** – Jetmaster Fires: www.jetmaster.co.uk; **Quartz floor and wall tiles** – Bristol Marble Company: 0117 9656 565; **Biotec sewage treatment plant** – Titan Pollution Control: 01264 353222 www.titanpc.co.uk; **Glass back panel** – Roman Glass Ltd: 01225 337433

CONVERTING A 17TH CENTURY THATCHED BARN

RURAL RENAISSANCE

Serial self-builders Chris and Trish Sale had a go at a barn conversion – with immensely satisfying results..

WORDS: JUDE WEBLEY PHOTOGRAPHY: NIGEL RIGDEN

"SELFBUILD? IT'S A Way of Life!" So says architect Chris Sale and his wife Trish who have now added a difficult but successful barn conversion to their three previous self-build projects. And, what's more, they are still smiling!

Chris and Trish built their first married home, in Barrington near Cambridge, in 1976; a contemporary style house with huge windows and double height rooms. Next they bought a plot in a smart Cambridge

"THE CONSERVATION OFFICER INITIALLY INSISTED ON LONG STRAW THATCH THAT HAD BEEN WETTED AND YEALMED ON SITE IN THE REAL OLD FASHIONED WAY."

The cart door opening has been filled with a new green oak framed structure, built with traditional mortice and tenon joints.

suburb much sought after by University Dons, built, and then retreated within a year of completion to the countryside, choked by the traffic and stifled by suburbia. The third project was "a lovely traditional home on a half acre plot back in the country," says Trish. "By this time we'd really got the hang of designing a floorplan that worked as a family home and only decided to move again when our daughter's interest in horses meant we needed more land."

When a derelict barn with planning and five acres came on the market in 1995 with an agent that Chris knew, they were tempted immediately. A farm next to a lovely village church only twenty minutes from Cambridge was being split into three lots: farmhouse, land and barn. "The barn was in a pretty terrible state and it took a lot of imagination to see its potential," explains Chris, "but we decided to go for it." They paid £150,000 but there were quite a few legal ➤

The floors are laid in new softwood 8" x 1" planks fixed to the chipboard below using cut nails and treated with Fiddes water based stain and wax.

Trish has been waiting to make the most of some beautiful brass door handles for years – great oak doors provided the perfect opportunity.

complications dividing the farm; they eventually got completion after about nine months.

There had been detailed planning consent for a conversion but Chris and Trish wanted a different design. The first proposal they submitted included a clay pantiled roof and an extra bay, with a link to the outbuildings. It was turned down, but a slight variation was later approved.

"Despite the fact that there are far more local barns with tiled rather than thatched roofs, the conservation officer insisted on thatch," recalls Chris. "Not only that, she initially insisted on long straw thatch that had been wetted and yealmed on site in the real old fashioned way. These days, modern thatchers do the wetting and yealming off-site and turn up with the thatch in bundles ready for the roof." In the end, the longer lived water reed was accepted due to the incompatible roof pitch but it was not until Chris had demanded an indemnity against failure of the type of roof they wanted that they backed down.

The barn was originally covered in sheeting and a great deal of imagination was required to see the full potential. After planning consent was finally granted, a structural engineer advised on the stability of the foundations and the process of renewing a partially rotting structure. Chris completed an outbuilding from which he was able to run his architectural practice but spent a lot of time on site supervising and helping. He employed a carpenter he knew who loved working on unusual projects and advertised for other labour, ending up with a couple of general builders and a young labourer employed on a day rate. Later, a bricklayer/plasterer was also taken on.

A huge central brick fireplace creates an informal partition between the main living space and the dining hall. ➤

The barn has two outbuildings, one of which (left) houses Chris's design studio.

"CHRIS HAS MADE SOME DOORS AND THE KITCHEN UNITS HIMSELF FROM OLD OAK FLOORBOARDS..."

The concrete floor and yard all had to be removed with the help of pneumatic hammers. Unfortunately, the old timber walls were rotten but the main aisle frame was in relatively good shape and survived largely intact. New foundations were dug in stages under the old frame, using a combination of a mini digger and hand digging. Meanwhile, the structure was held up with clamped timber supports and diagonal bracing. Expert repairs were carried out to the newly exposed feet of the main frame and some timbers replaced where the plates were rotten. The whole structure was then treated by Protim to protect it from fungal and insect damage.

Around the perimeter of the barn, a new plinth wall was built using reclaimed bricks and new timber frame external walls built, clad in treated timber weatherboard.

Inside, two bays of the barn have been given over to a spacious open plan living and dining area, divided by a huge brick chimney that houses a wonderful inglenook and open log fire basket. The space used by the cart doors has been filled with a new green oak structure into which has been inserted an arrangement of windows that Chris had made with softwood frames. Looking at this structure from the outside it is noticeable how well proportioned the window structure appears. "I'm really keen on getting the proportions of doors and windows right both structurally and aesthetically," explains Chris.

A discreet staircase leads off the main living area up to a galleried first floor landing above the dining hall. This occupies part of the central bay, along with two bedrooms. The floor level here is slightly higher than the bay either side, raised above the head of the cart door opening. Steps down to the bay either side – housing the master bedroom at one end and two further bedrooms and family bathroom to the other – add interest and give this barn conversion a very natural and homely feel.

Other little touches help create the rustic, country style interior. Chris has made some doors and the kitchen units himself from some old oak

The converted barn is almost unrecognisable from its former state when it was covered in corrugated metal sheeting.

floorboards and combined with traditional style fittings, these really look the part.

Chris and Trish have achieved excellent value for money on this really quite complex project and have ended up with an elegant yet homely property set in the beginnings of a clever architectural landscape with a vista cut through a copse of trees to a pedestal raised in the distance. The clever part is that the concrete from the old farmyard has been buried here to create the raised level that hides the road at the boundary of the property.

Chris and Trish have clearly derived pleasure from the fact that the conservation officer who caused them so much grief at the start of the project has since visited the completed project and praised the sensitivity with which it has been carried out. So will they build again? They certainly deny it for now – but don't bet against it! ■

FACT FILE

Costs as of May 2002
Name: Chris and Trish Sale
Profession: Architect and Housewife
Area: Cambridgeshire
House type: Barn Conversion
House size: 334m2 (incl. outbuildings)
Build route: Subcontractors and DIY
Construction: Oak frame barn + Modern timber frame outer walls
Finance: Private
Build time: Nine months (1997)
Land cost: £150,000
Build cost: £170,000 (inc outbuildings)
Total cost: £320,000
House value: £800,000
Cost/m2: £509

60%
COST SAVING

FLOORPLAN

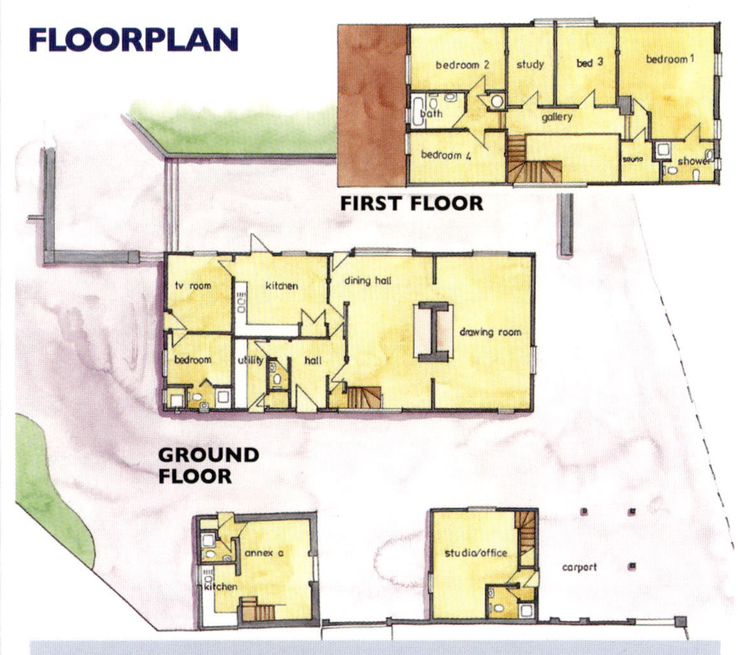

USEFUL CONTACTS

Architect – Rider Sale Associates: 01223 264436 (www.ridersale.net) **Structural Engineer** – Allan Baldry & Assoc: 01223 312784 **Master Thatcher** – C. Dyer: 01920 438403 **Thatching Advisory Service:** 01256 880828 **Insurance for Thatched Properties** – Thatch Owners Insurance Agency: 020 8506 6211 Country Insurance Services: 08457 660 063 NFU Mutual www.nfumutual.co.uk **Joinery** – W.J. Developments Ltd: 01223 837372 **Plastic Plumbing** – Hep2O: 01226 763561 **Reclaimed Materials** – Solopark: 01223 834663 **Reclaimed Bricks** – Brickability: 01656 645222 **Glass** – Solaglass: 01733 297800 **Beam & Block Floor** – Rackam House Floors: 01924 455876 **Clay Pantiles (outbuilding)** – William Blythe: 01652 632175 **Sanitaryware** – Ideal Standard: 01482 346461 **Timber Preservation** – Protim: 01223 234743

PERFECTLY BALANCED

Tim and Hazel Dawson have created a dramatic family home – packed with the latest hi-tech electronics – by converting a Grade II listed barn.

WORDS: JASON ORME PHOTOGRAPHY: JEREMY PHILLIPS

TIM AND HAZEL Dawson's Grade II listed converted barn – situated in a scenic valley to the east of Lancaster – dates from as far back as the 17th century but, today, visitors are greeted by a range of 21st century gadgets and gizmos that make the large (500m^2) property a pleasure to live in for the couple and their two young children.

Tim and Hazel had previously renovated houses and were looking for land to build a new home for themselves from scratch, but, with nothing on the market, their attention drew to the possibilities of converting a barn or other

The central triple height atrium space is punctuated by a galleried landing accessed by an elegant curved staircase.

Internal windows at third storey level help the Dawsons to retain the visual drama of the original structure while at the same time compartmentalising the space into a conventional modern home.

"I SPENT HOURS SOURCING GOOD DEALS ON THE INTERNET AND FOUND IT INCOMPARABLE..."

The family bathroom, situated off the first floor galleried landing, has an internal glass roof, ensuring as much light as possible enters the room, which has no external windows.

building. "We found one, fell in love with it, commissioned an architect and then, to our dismay, it fell through," says Tim. "We were devastated but our architect had heard about another barn that was going on the market and we decided to check it out."

The barn was part of a range of farm buildings that were being sold off in separate lots. "A developer picked up a large building in the development and this was the only one that was left," says Hazel. "I suspect that it wouldn't have appealed to a developer because it was so large and, unlike the rest of the surrounding buildings, was listed. We loved the fact that it was so open and the structure was so dramatic and tall inside."

The local council had other ideas. "It initially had permission for conversion into a restaurant," explains Tim. "They felt that the dramatic height space inside would best be served by having tables and a public space. But when we first looked around we could see that it would make a fantastic home."

Planning took almost two years to go through. "It was a very frustrating process," says Tim. "Initially we wanted to have three storeys but the fire ➤

regulations proved insurmountable and the conservation officer insisted on retaining as much of the original structure as possible, so after much wrangling we compromised with a mezzanine. The third storey bedroom would actually have been quite poky and we wanted to accentuate the original structure. The architect came up with the idea for a large 'church' window so you could see all the beams from the hall which tied in with what they required, as it turned out. We found dealing with them quite difficult – we kept asking them for positive input into the scheme but they only seemed willing to react to what we had come up with rather than be proactive. In addition, our building control officer's demands sometimes contradicted what the planners and conservation officer wanted." Hazel interjects: "We often found ourselves between a rock and a hard place."

Common sense eventually won, however, and Hazel and Tim, in true self-build tradition, bought a cheap caravan and parked it up on site. What commitment! "We can't exactly claim to have put ourselves through hell," Hazel backtracks quietly. "We were renting a home nearby and bought the caravan more as a site office and somewhere to stay if we were feeling brave! I was pregnant at the time and managed the project myself, as Tim was working full-time." This impressive arrangement was something that she fully enjoyed, as did Tim, who laboured on site where necessary and wired up a lot of the electrical systems. "I actually found tidying up to be one of the key jobs," sighs Tim. "Some of the subcontractors would invariably leave rubble in between the joists and wood shavings everywhere, which in addition to being untidy was a fire hazard."

But it is on the subject of the modern technology that Tim becomes really animated. Of all the rooms in the house, it is the small 'nerve

"WE'VE GOT A SPECTACULAR LOCATION AND A HOME THAT IS TRUE TO THE ORIGINAL FEELING OF THE PROPERTY, BUT DOESN'T PANDER TO IT."

centre' – a tiny operations room chocked full of patches, wires, pumps and, rather alarmingly, dozens of explanatory post-it notes – that Tim seems most keen on. "We installed a ground source heat pump," he explains. "It saves us, we calculate, around £1,200 a year compared to lpg. The payback time on it is just five years and we, so far, haven't had to use any form of back-up supply. Unfortunately we just missed the government grants now available. The heat pump powers the underfloor heating but due to the large amounts of solar gain from the central atrium and the vast amounts of stone, which act as a super thermal store, we use the heating system minimally." A heat recovery ventilation system helps to reduce the heat requirements even further.

The house is controlled by a home automation system, which enables the couple to adjust almost any of the intelligent functions of the home – the mood lighting, security systems, even the opening and closing of the curtains – from a phone anywhere in the world. In addition, a range of sensors are used to monitor the various temperatures and other vital statistics of the property across time – data for which, as yet, Tim has found no purpose other than purely interest. "It sounds terribly complicated," he smiles, "but future owners will be able to use it easily and enjoy the benefits." With industry experts predicting that homes ➤

"THE GROUND SOURCE HEAT PUMP SAVES AROUND £1,200 A YEAR COMPARED TO LPG..."

without home automation will become practically obsolete within a couple of generations, this 1715 vintage has clearly been brought right up to date.

The interiors in this spacious home are surprisingly cosy, with a high quality of finish applied – but not necessarily reflecting the budget. "I spent hours sourcing good deals on the internet and found it incomparable for prices," Tim says. "We initially budgeted £15,000 for the kitchen but only ended up with £6,000 to spend. But we found a fabulous kitchen complete with moulded stainless steel worktops to meet the price." In another bout of heady bargaining, Tim and Hazel decided to sell the rather dull grey flagstones that existed in the original barn and spent the money on some softer, more colourful Indian sandstone.

But it is the dramatic central atrium that greets visitors with that all-important 'wow' factor. A galleried landing runs around the first floor making this a great party space, while a delightful curved steel and oak staircase provides relief from the straight lines elsewhere. The triple height space is lit by a bank of rooflights and looks especially spectacular at night. In addition, an original cantilevered stone staircase – what the design mafia in London wouldn't give for one of those – is accentuated by a supporting steel structure and simple glass balustrading. "This was a particularly sensitive point with the conservation officer but I think that he was pleased with the contrast of new and old materials," says Tim.

"We've got a spectacular location and a home that is true to the original feeling of the property but doesn't pander to it," says Hazel. "It's interesting to compare what we've done to what the developer has done across the way and says a lot about the professionalism and dedication of our architect and builder." ■

FLOORPLAN

The layout is based around the triple height central atrium space. The five bedroom first floor includes plenty of flexible space, with bedroom storage built into the eaves.

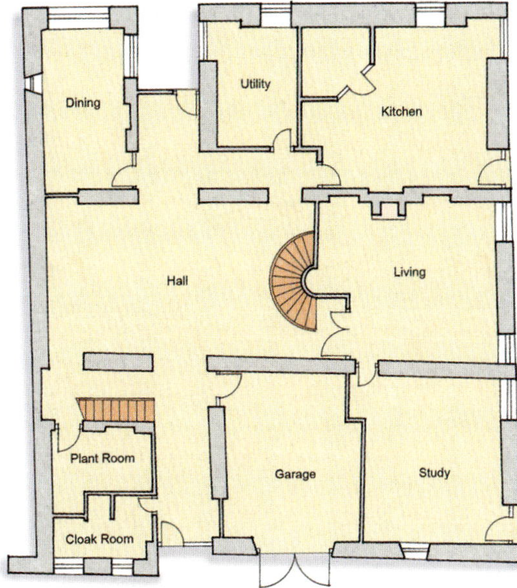

GROUND FLOOR

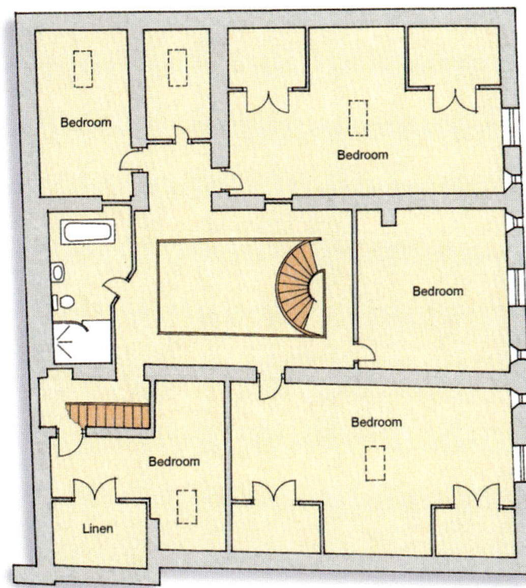

FIRST FLOOR

FACT FILE

Names: Tim and Hazel Dawson
Professions: Doctor and homemaker
Area: Lancashire
House type: Converted barn
House size: 500m^2
Build route: Selves as project managers
Construction: Masonry
Finance: Private
Build time: February '02 – March '04
Barn cost: £100,000
Build cost: £257,000
Total cost: £357,000
House value: £500,000
Cost/m^2: £514

29%
COST SAVING

CONVERTING BARNS

There are pros and cons in taking on an old agricultural building. In most instances – and as Tim and Hazel found – the problems of not being able to alter the exterior of the property and add new openings can be a barrier, but are offset by unique character, the sheer volume of space and a great location, often an isolated rural spot where a new house would never gain planning consent.

USEFUL CONTACTS

Architect – David Hacking, DH Design North West Ltd: 01524 272138; **Builder** – John Collis, Finchfive Ltd: 01524 734898; **Ground source heating and heat recovery ventilation** – John Cantor Heat Pumps Ltd: 01650 511575; **Underfloor heating** – David Robbens: 0800 454569; **Digital lighting** – Futronix Ltd: 01883 373333; **Light fittings** – Asco: 0870 120 1552; **Home automation** – Comfort Home Controls: 01244 680675; **Stainless steel worktops** – Universal Worktops: 01388 529243; **Kitchen** – Arena Kitchens: 01349 895014; **Glass bathroom roof** – Fast Glass: 01282 832948; **Metalwork and curved stairs** – Alan Dick Engineering Ltd: 01524 855011

smeg
ALESSI
ALESSI

FLOODED WITH LIGHT

Graham and Tig Chilton have converted their barn from an empty shell into a light-filled, contemporary family home.

WORDS: TONY GREENWAY PHOTOGRAPHY: DEBORAH EDWARDS

The Nolte kitchen is housed in a conservatory extension, which is a light and airy room with its glass roof and doors, and views of the river running directly outside the house.

GRAHAM CHILTON AND his wife, Tig, were looking for a house that was light, airy, modern and – most importantly – functional. "A lot of contemporary spaces these days look great," says Graham, "but you can't live in them. We wanted something that was unfussy without being stark – and usable, too."

A few years ago, Devon-based Graham was working as a joiner for a West Country property developer who owned a complex of 12 barns which were being sold for conversion. Graham and Tig bought a small corner barn for £65,000, a price which included all services to the building. The couple thought it was a space they could turn into their ideal home, and Graham had plans to do most of the conversion work himself.

"The barn we bought was just a shell," he says. "There was nothing in at all – no windows or doors. We had to subdivide the space and erect the party walls too. And just to be on the safe side, I re-tiled the roof."

In the entrance way to his part of the barn, Graham found two enormous pits that were eight feet deep. This was a former silo, where grain used to be stored, and would soon become his hallway. "The problem was, what to do with these enormous holes?" says Graham. "Then I thought: 'Hmmm... We could put a fish pond in there...'"

First, though, Graham and Tig sat down to design the look of the house from scratch. "I got an architect to help me," says Graham, "and we talked about what Tig and I had in mind. The architect came up with the basic ➤

"WE WANTED SOMETHING THAT WAS UNFUSSY WITHOUT BEING STARK – AND USABLE, TOO."

The bespoke stairs were made in white concrete by spiral staircase specialist Spiral Construction, and make a great viewing gallery for the fishpond. Graham and Tig opted to put green slate tiles on the floor and install decking (which the fish can swim under) directly by the front door.

design; but we already knew we liked the idea of a first floor accessed by a spiral staircase, and I wanted glass walls in the bedrooms. It had to be contemporary and bright."

As a carpenter Graham wasn't fazed by the work he was about to undertake; but even he underestimated the amount of time it would take to complete. "I think the most difficult part of the build was doing it all myself on evenings, weekends and the odd day off here and there," he says.

"At the time, I was working for the developer for 10 hours a day, then coming back to my own building site. It was like a busman's holiday. It took a year to complete, and I was so tired all the time." Yet being in the trade meant that Graham was in the enviable position of calling in a lot of favours from contacts: he knew a first-class tiler and plasterer/bricklayer to begin with. And because his father-in-law owned a second property nearby, Tig and Graham were able to live in that, rent free, while the building work went on.

Graham began by digging out the downstairs (and putting the resulting hardcore into those silo pits), laying a damp-proof membrane, fixing underfloor heating ("it cost about £2,000 – we bought the kit and just did it ourselves") and the steelwork for the upper level, and pouring a new insulated concrete floor. Constructing the pond – which originally ran across the entire width of the house – meant pouring a concrete base on top of the hardcore and then tiling the inside with glass mosaic tiles.

Building the first floor – which is home to two small bedrooms, a master bedroom with gallery en suite and a wet room, and is accessible via an ultra-modern white concrete spiral staircase – took Graham five evenings. But construction wasn't without its problems. ➤

"YOU DON'T NEED LOTS OF BRIGHT PAINT IN A PLACE LIKE THIS BECAUSE THERE'S COLOUR IN THE EXPOSED BRICK."

"I'd installed a Velux window in the back of the building, where we'd planned to put the en suite gallery bathroom," says Graham. "The original drawing showed that the en suite was going to be directly under it. But when we put the studwork up, we 'found' a big space on the opposite side of the master bedroom. It was ideal for the bathroom – but I'd already put the Velux in. So that was a bit of a cock-up."

Glass walls were included in the design of the two smaller bedrooms which overlook the fishpond, staircase, patio doors and dining area. "Having glass walls helps to create a sense of space because these rooms aren't particularly big," says Graham, revealing that the spare room (with its additional curved glass brick wall) is his favourite space in the house.

The most time-intensive part of the build – and possibly most expensive on a per square metre basis – was the wet room. "There aren't any windows in the wet room," says Graham, "so I put glass blocks into one wall to let borrowed light in from the next room. As a result, the wet room took the best part of a week and a half to tile because I had to cut round the glass mosaics individually and position them around the blocks. Plus, glass mosaic was expensive at the time – about £80 a metre. When I think about it, there was a big cost in the tiling and slating of this house."

Downstairs, with the external walls sandblasted, Graham made a feature of the exposed brickwork. "You don't need lots of bright paint in a place like this because there's colour in the exposed brick. Plus, if you put a light tone on the walls, the light reflects off it."

This luxuriously simple room makes a feature of the exposed brickwork, glass bricks in the bed head and industrial radiator. The staircase up to the en suite bathroom was constructed by Capricorn Engineering and cost £200.

The fireplace is fully functional and was designed by Graham and the builder out of block, vermiculite and concrete The exposed brickwork is a reminder of the building's heritage and helps to give the property its character.

Internally, screen walls have been used to create informal partitions between the living area, hallway and dining areas. "In the lounge area, we only have one small window on the back wall," says Graham. "But at the front we have these huge windows, so the sub-dividers allow the light to flood in."

Graham also built a fireplace in the lounge out of block, vermiculite and concrete, to form a material that's heat-proof and crack-proof. "There wasn't a drawing for the fireplace," says Graham. "The bricklayer and I just made the design up as we went along. We liked the look of the stainless steel Selkirk flue which runs up the master bedroom wall; so we left that exposed."

Two years ago, Graham added an extension to the side of the house with a wall divide creating a further two rooms. One of these spaces is now a playroom for the Chiltons' three-year-old daughter, Syd; the other became a glass-roofed kitchen area. "Originally, the kitchen took up part of the sitting room," says Graham. "It was OK, but it wasn't ideal, so adding the conservatory creates light; and putting the kitchen in the extension gives us more space."

At the end of construction, Graham mounted a grain storage pulley on the party wall as a reminder of the building's history. And, while he's now gone into the property-developing business full time and is currently busy working on another house, this one remains his first love. "One of the things I really like is that the whole building is flooded with light," he says. "It's been great to see the house develop over time and I hope we'll be here for a few years yet." ■

FACT FILE

Names: Graham and Tig Chilton
Professions: Property developer and editor
Area: Devon
House type: Barn conversion
House size: 139m²
Build route: Self-managed with architect and subcontractors
Warranty: None
Finance: Halifax
Build time: 12 months
Building cost: £65,000
Build cost: £80,000
Total cost: £145,000
House value: £350,000
Cost/m²: £576

59%
COST SAVING

FLOORPLAN

The ground floor area is almost entirely open plan, with the first floor housing three bedrooms and one bathroom.

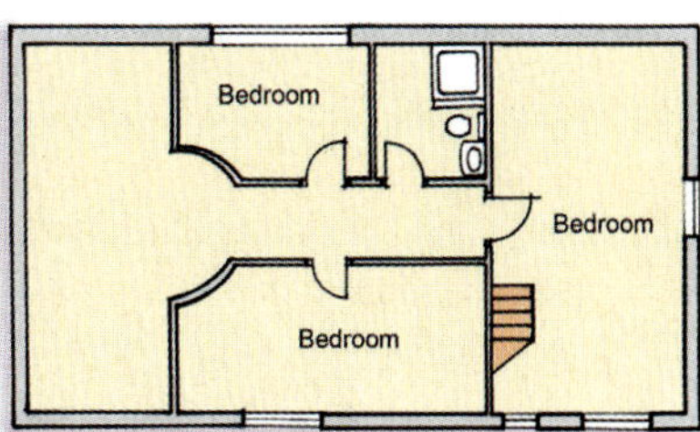

FIRST FLOOR

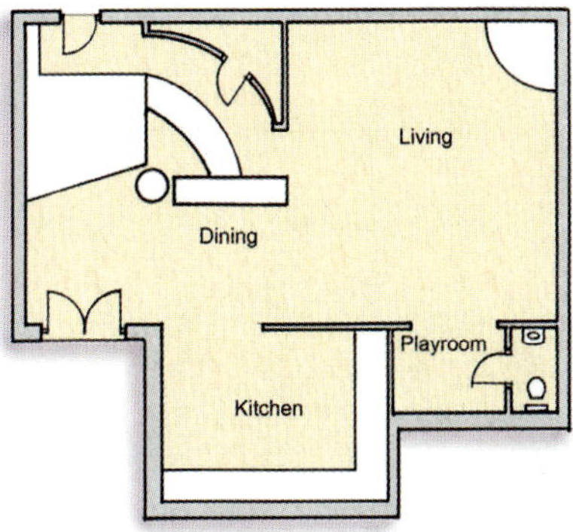

GROUND FLOOR

BARN CONVERSIONS AND VAT

Work involved in creating a new dwelling through the conversion of an existing building – such as a barn – is largely free of VAT. VAT-registered builders must charge the reduced rate of 5% on the supply of eligible materials and labour which you can then later reclaim on completion of the project, together with any standard rate VAT (17.5%) paid on materials bought yourself, under VAT Notice 719.

USEFUL CONTACTS

Kitchen – Nolte kitchens: 01279 868500; **Radiators** – Bisque Radiators: 020 7328 2225; **Plate metalwork** – Capricorn Engineering, Devon: 01392 841831, **Spiral staircase** – Spiral Construction: 01326 574497; **Slate floors** – United Tile: 01392 438116; **Light switches and sockets** – Gewiss: 01249 444734

CONVERTING AND EXTENDING TWO 17TH CENTURY BARNS

LINK TO THE PAST

Ian and Maureen George have created an impressive and spacious home by converting and extending a pair of run down Grade II Listed barns.

WORDS: MICHAEL HOLMES PHOTOGRAPHY: NIGEL RIGDEN

WHEN IAN AND Maureen George bought two run- down stone barns in South Oxfordshire, with planning permission for conversion to a house, they admit they gave little consideration to the implications of undertaking a major building project. "Had we thought too much about the ins and outs of what we were taking on, we might not have gone ahead!" admits Maureen.

The couple had lived next door to the pair of stone and timber Grade II Listed barns, which date back to the late 17th century, for almost twenty years and had long admired the chalky white rubblestone and brick walls. "So when our neighbour mentioned that he had noticed an estate agent measuring up and preparing sales details, we were immediately interested," recalls Maureen.

To help assess if such a project was financially viable, the couple consulted architect, Roger Danks, whom they had met through a mutual friend. Armed with an idea of the practicalities and the costs, plus an

The farmhouse style kitchen/ breakfast room is within the new-build link between the two barns. The floor is of reclaimed French terracotta. ➤

The white tiled pool is surrounded by white limestone and features a steam room shower enclosure and its own contemporary style, stainless steel kitchen. At the same time the couple built a new three-bay garage.

Three rubblestone walls and a tin roof were all that was left of a third barn that has now been rebuilt as a swimming pool enclosure.

estimate of the likely resale value via a local estate agent, they concluded that it would be worthwhile going ahead and managed to purchase the property and three acres of land at the asking price, before it went onto the open market. "In hindsight I suppose we were very lucky," admits Maureen.

The Georges felt that they could improve on the existing approved plans and so commissioned Danks to come up with some new design ideas. "The original design divided the smaller of the two barns into several small rooms, including three bedrooms upstairs," recalls Ian. "We felt that this completely wasted the principal attraction of the building – the space and the beamed roof. Instead we decided to keep this barn as a single living space, open to the rafters, and overlooked by a reading gallery."

The original design linked the two barns via a narrow passageway, to be built in place of an old lean to. Instead the Georges decided to build a whole new two storey section, giving them – in addition to a linking hallway – a kitchen/breakfast room and a large first floor gallery landing. "As the kitchen was new-build, we were able to incorporate lots of glazing, something that was limited elsewhere in the barn due to strict ➤

"THE CONSERVATION OFFICER WAS AGAINST THE USE OF RECLAIMED CLAY ROOF TILES BECAUSE HE FELT IT ENCOURAGES THE REMOVAL OF TILES FROM OTHER OLD BUILDINGS."

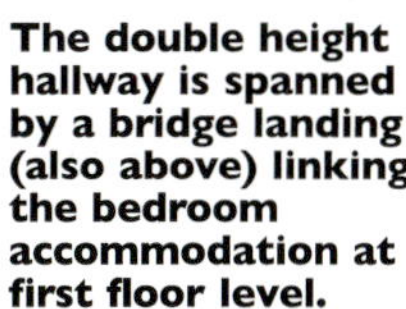

The double height hallway is spanned by a bridge landing (also above) linking the bedroom accommodation at first floor level.

planning policies. The views over the garden from here are lovely."

Despite the barns' listed status, consent was granted for the design alterations with little resistance, including a new two storey garage section. "The only issue we really had to push was over the use of reclaimed clay roof tiles," says Maureen. "The conservation officer was against their use because he felt it encourages the removal of tiles from other old buildings. Eventually he conceded though."

Re-roofing the barns was just one of many tasks necessary to bring the 17th century buildings up to a habitable standard. Covered for some years with only corrugated tin, the elements had found their way inside and consequently some of the timbers had started to decay. Several sections needed restoring and others required replacing altogether. All of the timbers then needed treating and preserving.

The solid rubblestone walls had survived in reasonable condition but, with no foundations to speak of, required underpinning and a damp proof course inserting. The original floors were simply earth and concrete and so these too required digging out and new reinforced concrete floors laid, insulated with extruded polystyrene, followed by a pipe in screed underfloor heating system from Wirsbo.

It took some time to find the right contractor for the project, but

The new section linking the barns provides a walkway from the dining area to the living room, plus a kitchen/ breakfast room.

when the Georges showed Tom Webster of Preservation in Action around the site, his enthusiasm for the project, and obvious knowledge of the key issues at play, instantly persuaded them that they had found the right man. This was further confirmed when they visited his previous projects and checked references.

Work began on site in November of 1995 under the supervision of Roger Danks who was retained on a percentage fee basis. Although there were working drawings, much of the internal detailing was decided on site. "Tom Webster was very good at coming up with ideas and he helped us choose a great many of the fixtures and fittings," says Maureen. "These include the design for the kitchen and the fireplace in the living room. He supplied most items, as he was able to invoice everything zero rated for VAT, under the rules for alterations to listed buildings."

The Georges wanted to leave as much of the oak roof timbers exposed as possible, so the roof was insulated with Rockwool bats, sitting on foil backed plasterboard fixed in between the rafters. These were covered on the outside with breathable membrane rather than traditional roofing felt, before being battened and tiled. The new sections of wall for the link and garage were built using solid masonry – 250mm aircrete blocks – clad with new stone laid to match the original walls. The timber ➤

The grounds include a walled vegetable garden and green house, an formal pond and lawns that gradually transform to wild flower meadows as they approach a newly created lake.

sections of the external walls were covered with foil backed plasterboard, leaving the timbers exposed on the inside of the building. Insulation was placed between new studs, followed by a breather membrane and new sawn oak boards.

There were very few problems on site, although one nasty surprise did arise early on when the newly dug out floor in one of the barns began to flood following a particularly wet week. The floor had been lowered in order to even out the levels between the two barns, and had gone too close to the water table. The only solution was to tank all of the floors up to DPC level at an additional cost of £12,000. "It was not a good moment, but it would have been far worse had we not found this out until after completion," comments Ian.

Such meticulous work, all of which had to be inspected by the conservation officer at regular intervals, inevitably took time and by April 1997, eighteen months into the project, the Georges were anxious to move in. "Our daughter was shortly due to give birth to our first grandchild and move into our old home," recalls Maureen. "Suddenly the pressure was on. We ended up moving in with the place not quite finished, although the living room, kitchen and bedroom were up and running."

The Georges gradually moved their furniture from their old home over the following year and brought on the gardens. Then in Spring 2000, they decided to convert a third barn on their site into a swimming pool enclosure. "All that was left of the barn were three walls, but that was enough to persuade the planners to allow us to rebuild it and put in a swimming pool," explains Ian.

"ALL THAT WAS LEFT OF THE BARN WERE THREE WALLS, BUT THAT WAS ENOUGH TO PERSUADE THE PLANNERS TO ALLOW US TO REBUILD IT AND PUT IN A SWIMMING POOL."

"This was probably the worst part of the whole project," recalls Maureen. "Having gone through nineteen months of building work, for it to then start all over again, destroying the new garden, was very difficult."

Once again the Georges called in Tom Webster who managed to turn three rubblestone walls into a magnificent swimming pool enclosure, complete with an impressive oak roof made up using timbers reclaimed from an old French barn. The three acres of gardens have since been relandscaped and they now provide a beautiful backdrop, complete with a lake and wild flower meadows.

"The views are fantastic, and we can enjoy the garden from both inside and out," says Ian. "Living here is like being on holiday all the time." ■

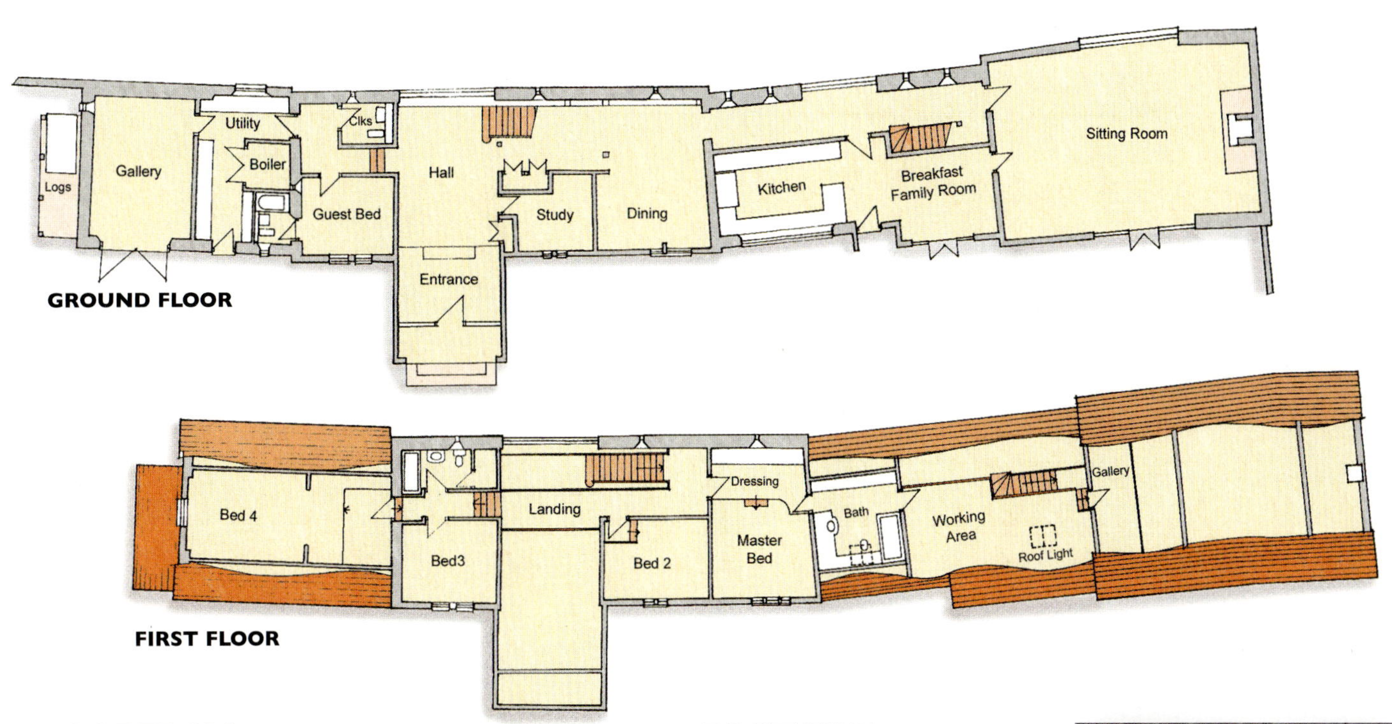

FLOORPLAN

The house includes four first floor bedrooms and three bathrooms, plus a ground floor bedroom with en suite for guests. The main barn and hallway are double height, overlooked by galleries. The link between the two barns and the garage/utility section are new-build.

FACT FILE

Costs as of Sept 2003
Name: Ian and Maureen George
Profession: IT consultant and lecturer
Area: South Oxfordshire
House type: Five bedroomed four bathroomed barn conversion.
House size: 500m^2 plus 275m^2 pool
Build route: Architect & contractor
Construction: Rubblestone walls, and oak framing.
Warranty: None
Sap rating: n/a
Finance: Private
Build time: Stage One: Nov '95 - May '97

Stage Two: April 2000 – Nov 2001
Average build cost: £1,500/m^2
House value: circa. £1,500,000

USEFUL CONTACTS

Architect - Roger Danks at Danks Reed Denby & Badnell; 01753 859880; **Contractor** - Websters 'Preservation in Action': 01502 478539; **Garden Designer** - Sarah Massey; 01491 612744 **Underfloor Heating -** Wirsbo (Uponor Housing Solutions): 01293 548512; **Breather Membrane -** Klober: 0800 783 3216; **Foil backed plasterboard** - Knauf: 0800 731 2108; **Insulation** - Rockwool: 01656 862621; **Kitchen Range** - Aga-Rayburn: 01952 642000; **Conservation Rooflights** - The Metal Window Company: 01993 830613; **Ceramic Tiles** - Fired Earth: 01295 812088; **Obscured Glass** - Pilkington Glass: 0121 326 5300; **Fireclay Sinks** - Brass & Traditional Sinks: 01291 650738; **Swimming Pool Design & Equipment** - Pool Design: 01666 840065

ROBERT LANG

CONVERTING A FORMER AGRICULTURAL BUILDING FOR £300/M²

AGAINST THE GRAIN

Denise and Mark Bigwood have created a spacious and elegant new home by converting two large stone granaries on a largely DIY basis.

WORDS:DEBBIE JEFFERY PHOTOGRAPHY: NIGEL RIGDEN

WHEN I FIRST told Denise that I intended to design our new home in a style similar to that of a Pizza Express restaurant she burst into tears," recalls Mark Bigwood. The couple were sitting in one of the modern pizza parlours at the time, and Mark realised that such slick finishes, minimalist lines and stainless steel fittings would suit the large stone granaries that he and Denise had purchased to convert. "Initially Denise had dreamed of a traditional farmhouse and was terribly upset, but she soon came around to the idea," Mark continues. "In fact, she's quite an expert on the subject of minimalism now!"

The Bigwoods had previously converted an old school in Wales, but decided in 1999 that they wanted to move on. Friends obtained the sale catalogue for two large stone built barns with an attached asbestos shed and two single storey outbuildings, which had formed part of a large Victorian country house estate near Bristol. "We were out of the country at the time and only had three weeks in which to organise our finances and submit a sealed bid," says Denise. "Lord Wraxall, who owned the Tyntesfield Estate, was selling seven properties – but we were only interested in Higher Farm Granary,

The kitchen units (below) came from Magnet and were adapted and personalised to suit the contemporary style of the interiors.

which had been a working granary until fairly recently."

Mark and Denise made a bid of £225,000, which secured the buildings which were full of silos and old machinery. Although they are set in beautiful countryside, the structures have a distinctly industrial feel, which set the tone for their conversion. "Two of the main buildings are stone with oak trussed roofs, whilst the third was an asbestos-clad farm building – attached to the larger of the two stone granaries," explains Mark. "They were akin to industrial warehouses, and we decided to convert them in the style of Docklands apartments and New York lofts rather than traditional barns. Denise took a little persuading but agreed that, as ➤

"THEY WERE AKIN TO INDUSTRIAL WAREHOUSES, AND WE DECIDED TO CONVERT THEM IN THE STYLE OF DOCKLANDS APARTMENTS AND NEW YORK LOFTS RATHER THAN TRADITIONAL BARNS."

The kitchen units came from Magnet and were adapted and personalised to suit the contemporary style of the interiors.

long as she could still have an Aga, we should pursue the idea of uncluttered contemporary interiors."

Although the property came with planning permission, Mark sketched some significant amendments to the original design, employing the services of an architect friend to draw up the revised plans. The two large granary buildings have been linked by a single storey porch to create a single five bedroom house. The asbestos farm building has been rebuilt as a timber framed Amish-style building, clad in timber, housing a two bedroom self-contained annexe above a workshop and garage, whilst

"WITH A BUDGET OF JUST £150,000 THE BIGWOODS' FUNDS WERE EXTREMELY TIGHT..."

the two animal sheds are now stables currently used for storage.

Remarkably, the planners accepted everything without question as, with the exception of the joinery, the exterior elevations of the buildings remain virtually unchanged on three sides, and are totally in keeping with their farmyard environment. The very private south aspect has been fitted with large windows which flood the interiors with light. "It was ➤

A gallery overlooks the sitting room in the larger granary, which features a contemporary fireplace designed by Mark.

"AS GRANARIES THEY HAD BEEN CONSTRUCTED WITH THICK WALLS OF LOCAL LIMESTONE TO KEEP THE GRAIN DRY."

A gallery allows you to look down into the sitting room, which features a contemporary stainless steel fireplace designed by Mark.

Denise's idea to extend one of the windows from floor to ceiling, which almost cuts the middle granary in half," says Mark.

The work involved to bring about such a transformation was daunting, but the Bigwoods were determined to get involved with as much of the hands-on building as they could in order to retain a degree of control and to keep costs down. Other than plastering and second fix wiring, Mark and Denise undertook everything from the carpentry to the landscaping – with the help of their son and Mark's long term colleague, Phil Francis.

Not content with a single project, they also purchased a run-down house, which they quickly renovated to provide a place to live whilst converting the smaller of the two granaries. Once it was habitable the couple moved in, living in their new kitchen and master bedroom until the rest of the conversion was complete.

"The buildings were actually in excellent condition," says Denise. "As granaries they had been constructed with thick walls of local limestone to keep the grain dry, and had recently been re-roofed with clay pantiles. Although we decided to reduce the levels of the ground around the outside in some places we have had no problems with damp. We needed to underpin one corner of the smaller stone granary but, other than that, the only structural problems were caused by rats which had chewed through some of the timbers. One of the oak trusses had to be braced where it had been gnawed dangerously thin."

The couple bought their own three tonne mini digger, which became an essential piece of kit – enabling them to tackle most of the excavation work and lay the drains. "Phil and I used it to lift the heavy steel joists into position for the first floors," says Mark, a heating engineer. "It cost us £3,250 and, once we had finished, we sold it on for £2,250."

With a budget of just £150,000 to complete the 500m^2 building, the Bigwoods' funds were extremely tight, with low-cost mass-produced products specified wherever possible. "Most of the doors and windows are standard sizes – we just chose the best quality hardwood on offer,"

says Denise. "Any bespoke joinery came from a local company, which matched it to the other units."

The building is relatively energy efficient thanks to the thick stone walls, high levels of insulation, low emissivity glass and underfloor heating. Window openings to the north and east façades have been kept to a minimum, with large expanses of glazing to the south and west maximising the passive solar gain.

"The larger granary is the most impressive building, with its high vaulted ceilings," says Denise. "The bedrooms and bathrooms have been built in modules which sit between the existing oak trusses – they actually appear to be floating." There is a two storey bedroom with an integrated staircase taking you up to the top floor. A gallery allows you to look down into the sitting room, which features a contemporary stainless steel fireplace designed by Mark. Even the coffee table was made by the Bigwoods, who have considered every last detail of their new home.

"We were lucky enough to source most of our materials locally," says Mark. "Even the stainless steel tension wires for the gallery and stairs were found nearby. Lighting was one of the most important considerations. We wanted to make sure that, although the house is decorated in white, we could introduce warmth and colour with lighting effects both inside and out. In the summer the rooms are bright and airy whilst, during the winter, the underfloor heating and thick stone walls create a cosy atmosphere. We have enjoyed the conversion so much that we may consider a new build in the future. Very modern, of course!" ■

FLOORPLAN

The two stone granary buildings have been converted into a five bedroom house, joined by a single storey porch. The smaller granary contains an open plan kitchen and snug, with the master bedroom suite above and an additional two bedroom self-contained annexe above the workshop and garage.

FACT FILE

Names: Mark and Denise Bigwood
Professions: Heating engineer and business analyst
Area: Bristol
House type: 7 bed converted granary
House size: 505m^2 + stables
Build route: Selves and contractors
Construction: Limestone walls, pantile roofs
SAP rating: 86
Finance: Private +Britannia
Build time: 26 months
Property cost: £265,000
Build cost: £151,917
Total cost: £416,917
House value: £1.25m
Cost/m^2: £301

67%
COST SAVING

Cost Breakdown:

Fees	£2,479
Electrics	£8,327
Plant and tool hire	£2,311
Drainage and guttering	£3,679
Labour	£34,650
Plumbing and bathrooms	£7,210
Heating	£5,160
Joinery	£16,344
Kitchen	£15,477
Annexe conversion	£13,632
Flooring	£7,821
Insulation	£7,428
Landscaping and pond	£8,154
First floor	£4,871
Sundry materials	£14,374
TOTAL	**£151,917**

USEFUL CONTACTS

Architect – Ian Bower: 01225 421671; **Structural engineer** – Alan Leonard Associates: 01179 215244; **Insulation** – Seconds Insulation: 01544 260501; **Steelwork** – Bristol Steel: 0117 9828131; **Plastering** – Ian Chapman: 01179 839440; **Underfloor heating** – Nuheat: 01404 549770; **Kitchen units, doors** – Magnet (Bristol): 0117 9507109; **Lighting and electrical** – Screwfix: 0500 414141; **Rainwater goods** – NRS Guttering: 01249 823160; **Bespoke windows** – Worlea Glass and Joinery: 01934 515470; **Carcassing, timber and machining** – William Staddon and Co.: 01275 873214; **Landscaping – Four Seasons: 01275 844716**

CONVERTING A BARN ON A BUDGET

SPLIT-LEVEL LIVING

Andy and Sandy Best have converted a barn and attached buildings into a characterful family home.

WORDS: CLIVE FEWINS PHOTOGRAPHY: ROB JUDGES

Once the unlisted barn was examined carefully it was clear that only small sections of the original frame could be saved It has been turned into a new home with traditional style interiors.

ANDY BEST TRACES his passion for split-level living right back to his youth in South-West London, where the terraced family home he lived in was built on the side of a former gravel pit. "Because of the unusual shape of the ground, the house was built on four storeys, with different levels all over the place," says Andy.

"My grandparents lived in a house with an equally unusual layout. It had a semi-basement and a half flight of stairs at the front that led into a passageway with rooms leading off and steps out of the scullery to the rear garden, so I suppose all this lodged itself deep in my subconscious."

Forty years later, Andy and his wife Sandy have just completed a house built along similar lines, except that it blends traditional and reclaimed materials with a more modern interpretation of split-level living and welds together three separate buildings – two new ones and one old one, a rebuilt 17th century barn.

"All my life I have lived in unusual houses, and this one – we reckon that between us we have undertaken over 80 per cent of the work – is no exception," says Andy. "Soon after we married in 1974, we moved out of London to Suffolk and purchased three derelict cottages. Sandy worked as an administrator for a local cabinet making company and I spent the winter renovating the cabinets and the summer doing building maintenance work for the diocese of St. Edmundsbury.

"We divided the cottages into two units. Both had cellars and were built on several different levels, and by the time we had renovated them

"BECAUSE OF THE UNUSUAL SHAPE OF THE GROUND, THE HOUSE WAS BUILT ON FOUR STOREYS, WITH DIFFERENT LEVELS ALL OVER THE PLACE."

➤

The rebuilt barn retains a large section of clay lump wall that forms an interesting vernacular feature, together with a fireplace created from reclaimed bricks.

and divided the proceeds into two with our partner in the project, we had enough money – £40,000 – to purchase the three-and-a-half acre plot in our present village. It contained a cottage in need of renovation and two derelict agricultural buildings – one of them the barn that became our house."

Andy and Sandy spent the next five years converting and renovating the cottage, then in 1989 moved out into a caravan on their land, where they lived for three years while planning the next stage of the development of the plot – rebuilding the barn, demolishing a small outbuilding and constructing a separate wing on the site, linking the two by means of a kitchen.

"We knew it would be a tricky and challenging project, but we could see that the idea of linking the three wings of the house at separate levels had the potential to create a really interesting home," Andy explains. Sandy continues: "Andy has a very good eye and we are both very determined, so although we knew there would be opposition, we felt we had the experience and resolve to proceed with it."

However, they had no idea the planning process would be so lengthy.

Extra space was achieved by digging down thus creating another level.

"We never found out where the opposition came from but there was a lot of it!" says Andy. "We were also not helped in our application by the fact that someone had recently converted an adjoining building into a house without planning permission, and we believe a lot of local people were against us because of this."

Their plans for the house were turned down six times by Breckland District Council, mainly because planners ruled that it was outside the village settlement boundary, although the plan was for the house to share its rear wall with the line that marked the outside edge of the boundary.

"The planners were clearly going to stick rigidly to these guidelines and we eventually employed a planning consultant, Geoffrey Lane," explains Andy. "At appeal we were able to show that two previous schemes in the village had not stuck strictly to these guidelines." Once they had won their appeal, Andy and Sandy had no problem getting the scheme – they had prepared a seventh version by this stage – through the planning process.

"We made a bad mistake when we bought the site in 1984. We were told then by local planners that there should be no problem with the scheme we had in mind. But we failed to ask for confirmation of this in writing and seven years later we found that there had been a big

"WE WERE ALSO NOT HELPED IN OUR APPLICATION BY THE FACT THAT SOMEONE HAD RECENTLY CONVERTED AN ADJOINING BUILDING INTO A HOUSE WITHOUT PLANNING PERMISSION."

clamp down on barn conversions. Really it was very bad timing — and that was our fault. Fortunately, using Geoffrey – he is a former planning officer – was the key to everything. He made it all possible."

Once the unlisted barn was examined carefully, it was clear that only small sections of the original frame could be saved. Andy had re-roofed much of it with reclaimed pantiles soon after they arrived, but the roof eventually had to come off completely and be replaced by these pantiles, plus a second batch he obtained locally.

"We were well established in the village by then, so we were able to source and buy many of our materials locally," says Andy. This proved very fortuitous as they had a budget of £57,000 from the sale of the previous cottage and £15,000 in interest and savings and did not wish to take out a loan.

With such a small budget for what amounted to a total rebuild of a 200m^2 house, and with Andy working elsewhere all summer, they knew it would be a long project.

After three years in the caravan, they were able to move on to phase one of the new house. This is the area at the far end they call the 'big sitting room'. It measures four-and-a-half by nine metres and stands on the site of a former animal shed that had to be completely demolished. It has a mezzanine sleeping area that sits above a basement containing two bedrooms.

"I realised that to get the roof height correct and to keep the big full-height room we wanted, the only way to gain more bedrooms was to go down," explains Andy. "It meant more work, but in a sense the site dictated this move. And of course it gave us yet another level and added interest to the house." The geology – chalk – was ideal for the excavation, which took place in a day."

The 'big sitting room' is linked to the rebuilt barn – which is at a higher level – by the new single storey kitchen. The rebuilt barn retains a large section of clay lump wall that forms an interesting vernacular ➤

The 'big sitting room' is linked to the rebuilt barn – which is at a higher level – by the new single storey kitchen.

"THE WALLS ON THREE SIDES ARE ENTIRELY NEW – MADE FROM BLOCKWORK WITH EXTERNAL INSULATION BENEATH THE BLACK HORIZONTAL WEATHERBOARDING."

feature, together with a fireplace created from reclaimed bricks.

Both give this section of the house, which has a staircase rising to the master bedroom on a mezzanine above, a rugged, traditional feel. Although the roof is new, Andy managed to retain the oak wall plate, some braces and the main corner posts and tie beams of the original oak frame.

"It is enough of the original building to give this rebuilt section of the house lots of character, but the walls on three sides are entirely new – made from blockwork with external insulation beneath the black horizontal weatherboarding," says Andy. A bank of double glazed windows with low-e glass faces the large garden and forms much of the south wall.

Because of their low budget, Andy and Sandy used a lot of reclaimed materials, many of which they managed to obtain free or on a barter basis. This, together with a high component of DIY labour, meant that it was nine years before Andy and Sandy finally declared the house finished. "We reckoned in all we probably spent £70,000 – it just shows what you can achieve on a low budget if you are prepared to put in a lot of effort yourselves. The result is a very interesting house built on no less than seven different levels!" ■

FLOORPLAN

BASEMENT FLOOR

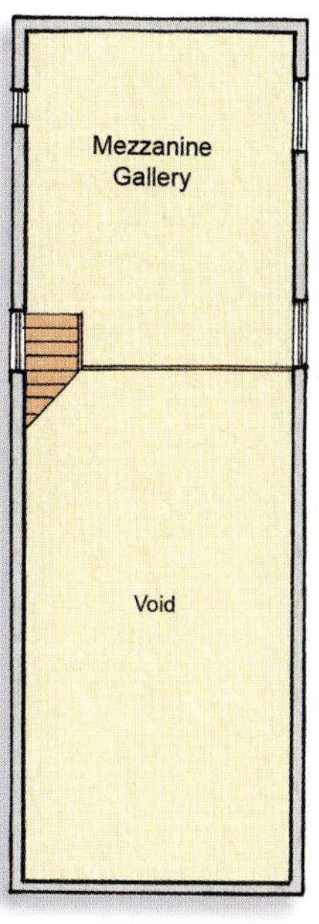

MEZZANINE LEVEL

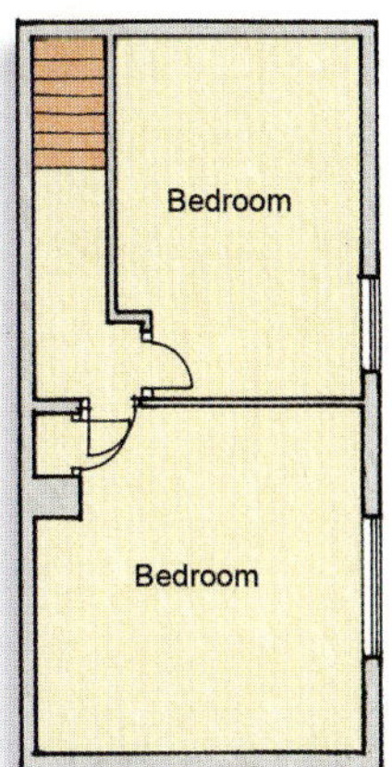

While a large ground floor houses most of the living accommodation, the Bests have incorporate mezzanine levels to introduce auxillary sleeping and quiet space, in addition to a basement containing two bedrooms.

BARN CONVERSION DESIGN

Barn conversions are a great way of creating a new home in a part of the countryside where new developments would otherwise be unacceptable from a planning point of view. For more information see Homebuilding & Renovating magazine each month with more barn conversion stories and inspirational features, renovation tips and technical know-how

USEFUL CONTACTS

Design – Gary Ward: 01953 688128; **Planning consultant** – Geoffrey Lane: 01603 219487; **Windows and external doors** – Tradspec: 01603 279834; **Stairs and internal doors** – Doug Bradley: 01842 753580; **Electrics** – John Brady: 01953 483784; **Plumbing and heating** – John Meadows: now retired; **Antique pine and other reclaimed building materials** – Womack Reclamation: 01953 860566; **Reclaimed bricks** – Aspect: 01953 717777

FACT FILE

Names: Andy and Sandy Best
Professions: Handyman and company administrator
Area: Norfolk
House type: Four bedroom detached
House size: 208m²
Build route: Selves as main contractor
Construction: Part oak frame, dual skin blockwork and brickwork, external weatherboarding
Warranty: Architect's Certificate
Finance: Private
Build time: Sep '92 – April '03
Land cost: Already owned
Build cost: £70,000
Total cost: £70,000
House value: £465,000
Cost/m²: £334

85%
COST SAVING

Cost Breakdown:

Groundworks and site preparation	£1,500
Excavation of basement, foundations and reinforcement	£4,500
Fee, including planning consultant and appeal	£3,500
Other fees	£1,600
Windows and doors	£8,000
Heating and plumbing	£3,000
Kitchen	£2,000
Sanitaryware	£4,000
Reclaimed timber	£3,000
New roofing and other timber	£10,000
Other reclaimed materials	£5,000
Labour	£5,000
Miscellaneous	£10,400
TOTAL	**£69,500**

There was already plenty of height in the structure. Geoff used it to add drama to the design.

BEST OF BOTH WOLDS

Geoff and Margaret Evans have converted an ancient stone barn into an exceptional modern home, with a brilliant mix of the traditional structure and contemporary design.

WORDS: JUDE WEBLEY PHOTOGRAPHY: MARK WELSH & WALTER DIRKS

The main living space is simply astonishing – three storeys tall and twelve metres high.

CONVERTING THE LARGEST tithe barn in the Bath and North Somerset area into a stunning contemporary home provided designer Geoff Evans and his wife, Margaret, with an opportunity to use their full range of design and project management skills. The Japanese style courtyard and a charming English country garden provide the perfect exterior complement to this most handsome and stylish project.

Perhaps surprisingly, the whole thing has been carried out with amazing economy, given the lack of compromise on the quality of the materials used. English oak joinery and kitchen, cast aluminium rainwater goods, stainless steel door furniture, Philippe Starck bathroom fittings, Vola taps, generously specified levels of insulation, lavish standards of landscaping and planting, along with a great deal of minute care and attention in the detailing — this is no low budget production. The site cost £116,000 in June 1996 and the conversion was completed for under £215,000, including landscaping, planting and conversion of the byre into a four car garage, studio, workshop and garden store. All in all, total costs amounted to £329,000. It's a lot of money but it's not without its rewards – 390m^2 of internal space in the house and 110m2 in the byre/garage. The property has been valued at well over £1 million.

"WE MADE NO ATTEMPT TO MIMIC THE OLD – OUR DRIVING MANTRA WAS TO ILLUSTRATE THE DIFFERENCE..."

➤

The kitchen was handmade in oak by David Oldfield. The stone worktop came from Kirkstone Quarries and the stainless steel surfaces from BM Stainless.

The external walls around the main entrance at both front and back were altered to allow an alcove, sheltered from the elements with the new stone-work, cut into the existing courses seamlessly.

Geoff's design philosophy guided the project. "The mistake people make with barn conversions is that they try and turn them into ordinary homes using domestic type fittings," he says. "With our barn, the whole structure was one enormous space. We wanted to create a contemporary dwelling and keep the idea of space, volume, light and openness intact."

"Why buy a barn just to turn it into a three bedroom semi? Why chop up all the space rather than make a feature of it? I wanted to maintain a sharp contrast between old and new. We made no attempt to mimic the old – our driving mantra was to illustrate the difference." The idea for the interior was to create separate areas but let one flow into the next, with as few internal doors as possible. There was plenty of height in the structure used to create drama in the design. There was also a desire to retain the barn-like, non domestic appearance of the external structure.

"The original planning consent design had three staircases, which divided the house into three separate parts – you wouldn't have been able to get across the whole building," explains Geoff, failing to disguise his frustration at such lack of vision. "There were more holes through the structure for new windows, which made it more domestic. In reality it was nothing more than a scheme to get planning permission, which is often what you get when you buy a barn or building plot with planning."

The first stage of the design process was to establish a fixed starting point for the floor levels. The Evans' wanted at least three floors. These needed to relate to the timber structure at the top of the building, part of ➤

The simple modern staircase serves as a central feature and manages to work very well amongst the old oak beams.

which would form the top floor. A process of measuring down from this point established the floor levels for the two lower floors. To give proper room heights, the lowest part of the house needed to be excavated below what turned out to be the foundations of the outside walls, which then had to be underpinned.

A structural engineer was retained to work out any necessary additional bracing. "Although I could have worked it out myself, I decided to use a well known local firm on the basis that the building inspector would be impressed and not give me too much trouble," acknowledges Geoff.

Spaces for different activities were then identified, with the centrepiece being the astonishing main living space. Over three storeys tall and 12m high, this is truly an impressive space. "I'm really pleased with the sculptural way the highly contemporary staircase works with the old barn beams," says Geoff. "Both Margaret and I work from home and the design incorporates working spaces for us both. We've also been able to include a large games room and a two storey bedroom for our son, Martyn."

Setting out each phase was regarded as vitally important. For instance, the lines of the oak floorboards in the main sitting room had to align properly with the Derbyshire limestone floor in the hall area, which in turn had to align with the door opening. Outside, the paving had to align with the centre of the front door and so on. All of these details were supervised personally.

The external walls around both main entrances were altered to allow an alcove – effectively a porch – sheltered from the elements, with the new stonework cut into the existing courses to avoid an unsightly join. Geoff designed the doors himself, taking great care with how each English ➤

"'WE SUPPLIED EVERYTHING THAT I WOULD CALL A 'FINISH'... IT'S POSSIBLE TO SAVE 20-30 PER CENT BY WORKING THIS WAY..."

oak board fits into the overall plan. He has also decided to leave out the standard threshold in the door frame, preferring the door to be flush to the floor protected only by draught stripping.

The practicalities provided some difficult choices. There were problems with the original contract to purchase the barn, which had been advertised as having utilities on site. "They weren't on site," recalls Geoff. "In fact we had to negotiate with the farmer, agreeing to install a pipe into his barn next door in exchange for an easement across his land. It's impossible to over emphasise the importance of utilities and access rights. If you buy a house it's likely that these things are already sorted out. A barn may only have rights of agricultural access – domestic access may not have been negotiated. Most people believe what's on the sales documents – I don't. I need to physically see the pipe sticking out of the ground."

Having finally exchanged contracts, a new planning application was submitted and approved with no great drama. Work began on site before the consent was received. Notice was served on the local authority that the work being carried out was not at variance with the original consent. So the utilities were brought in, boundary walls rebuilt and repointing undertaken. Also, the soakaway for the septic tank was installed, soil was brought in and the main lawn was seeded. In fact, the whole of the side garden was in place and planted up before work had started in earnest on the inside.

Geoffrey acted as project manager throughout the project, while ➤

"THE BARN HAS ATTRACTED A COACH TOUR OF ELECTED COUNCILLORS WHO CAME TO SEE HOW BARNS SHOULD BE CONVERTED..."

Margaret did the research, ordered materials and ran the finances. One main contractor was used to carry out all the main structural work and each package of specialist work was put out to tender from subcontractors, who were encouraged to supply their core materials themselves. "We supplied everything that I would call a 'finish'," says Geoff. "I reckon it's possible to save 20-30 per cent of the total costs by working this way. To start with, a contractor will mark up subcontractors' costs by 10%. Secondly, he will use established firms who are safe, paying 10-20% more than is necessary. Thirdly, there's the mark up on expensive materials. As long as you're prepared to work with the subbies and make sure the materials are there on site when they're needed, I'd recommend this way of working."

Geoffrey feels it was important to get professionals in at important stages. "People seem to resent spending money on architects," he reflects. "The attitude tends to be, 'why should I pay an architect 5-10 grand, when someone down the road will do the drawings for £1,000'. Yet they'll pay loads for a nice sofa or other physical products. Similarly, they won't spend the money on gardens – a great shame, since the setting for the house is an important part of the whole."

"The setting for the house is an important part of the whole," says Geoff, who designed the landscaping himself.

The conversion is already well known locally and has attracted many visitors to both the house and gardens – including a coach tour of elected representatives from the local council, who came to see how barns should be converted. By combining a modern style with the traditional character of the original building, the Evans' have achieved a benchmark conversion. ■

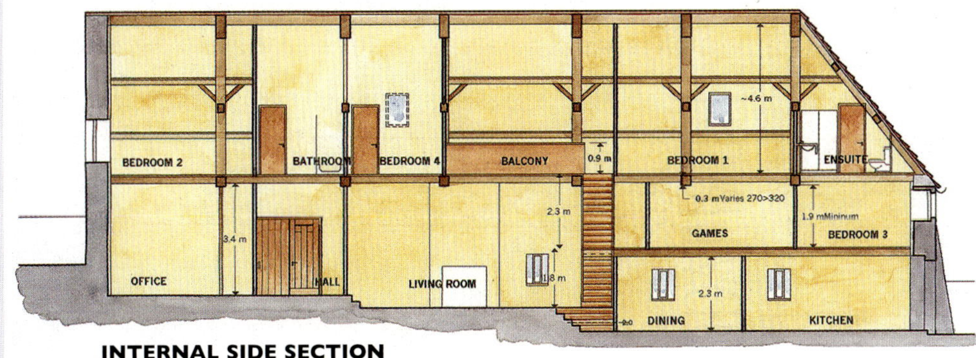

INTERNAL SIDE SECTION

FLOORPLAN

The central section of the main living area is open to the ridge. The three storey section was created through excavation.

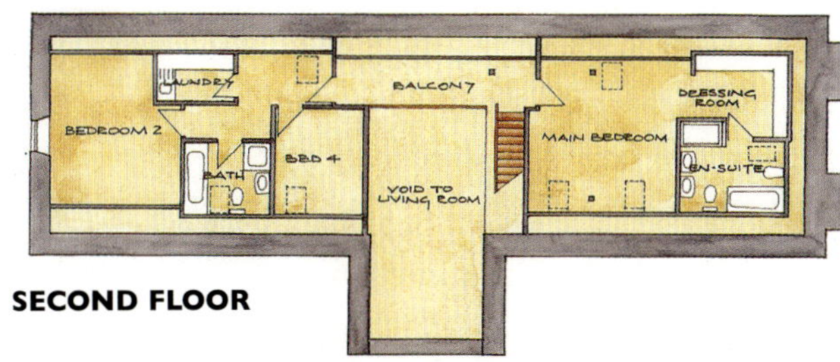

SECOND FLOOR

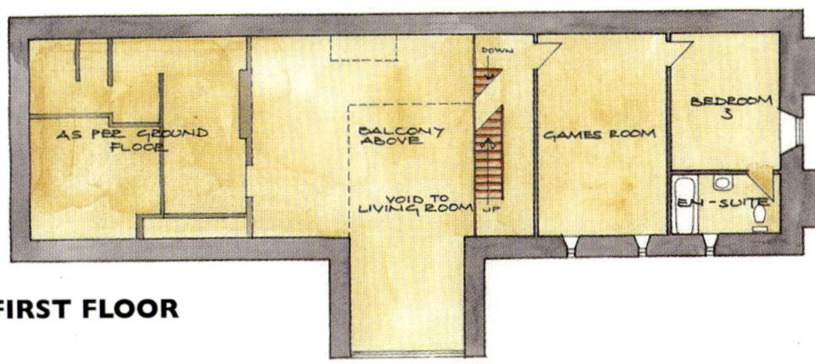

FIRST FLOOR

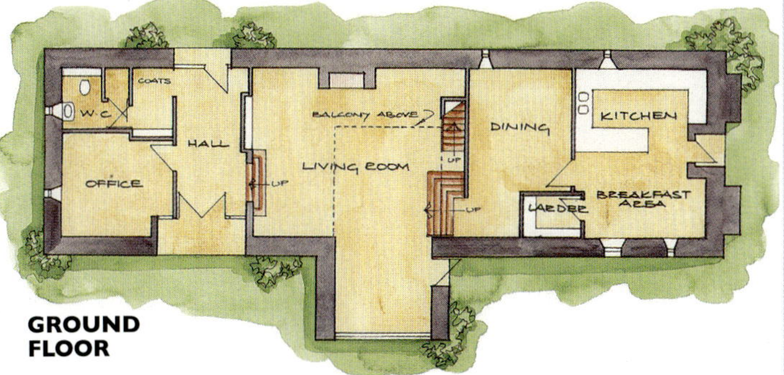

GROUND FLOOR

FACT FILE

Costs as of Jan 2000
Name: Geoffrey and Margaret Evans
Professions: Landscape Architect, Mother & Administrator in business
Area: Bath
House type: Stone Barn Conversion
House size: 390m² plus 110m² outbuilding/garage
Build route: Self project managed & subcontractors
Construction: Traditional
Warranty: None
Finance: Bank
Build time: Ten months
Land cost: £116,000
Build cost: £213,000
House value: £925,000
Cost/m²: £546

64%
COST SAVING

Cost Breakdown:

Professional fees	£4,000
Utilities	£4,500
Main building works	£75,500
Roof	£11,000
External oak joinery & glazing	£12,000
Internal doors & carpentry	£4,500
Electrics & fittings	£9,000
Plumbing/heating + fittings	£20,000
Kitchen & laundry furniture & fittings .	£15,000
Floor finishes	£10,500
Painting & decorating	£4,500
Outbuilding/garage	£10,000
Hard landscaping	£15,000
Planting	£2,500
Additional work	£15,000
TOTAL	**£213,000**

USEFUL CONTACTS

Main Contractor – Nigel Lloyd: 01373 473642; **Structural Engineer –** Mann Williams: 01225 464419; **Electrical Contractor** – Graham Newman: 01225 426377; **Glass & Glazing** – Bath Glass & Glazing: 01225 447042; **Windows & External Doors** – Woods of Wales: 01938 554789; County Hardwoods: 01823 443760 **Internal Doors (solid chipboard with oak veneer, fire door quality)** – Vicaima: 01793 532333; **Timber Floors** – Penrose Wood Industries: 01825 872025; **Derbyshire Limestone Floor** – Cadeby Quarry: 01709 866628; **Pipe Trace Heating** – Raychem: 01793 572663; Perimeter Heating – ICC: 01483 537000; **Architectural Hardware** – Webb Lloyd: 0121 622 1798 bought from TBKS: 01225 462090; **Handmade Kitchen** – David Oldfield: 01225 777325; **Kitchen Stone Worktop –** Kirkstone Quarries: 01539 433296; **Kitchen Stainless Steel Surfaces – BM** Stainless: 01937 838000; **Kitchen Taps** – Vola: 01525 841155 **Sanitaryware** – Philippe Starck, bought from CP Hart: 020 7902 100 Ideal Standard: 01482 346461; **Paving Slabs** (Perfecta Buff with polished surface) – Marshalls: 01422 306000; **Trees, Shrubs & Plants** – bought wholesale from Wyevale Nurseries: 01432 352255

CONVERSION OF AN OAK FRAMED BARN

BARN AGAIN

Vic and Sue Wagland have converted a dilapidated oak-framed barn into a stylish family home..

WORDS: DEBBIE JEFFERY PHOTOGRAPHY: JUSTIN PAUL

VIC WAGLAND AND his wife, Sue, owned a Grade II listed farmhouse with a number of ramshackle outbuildings, and had often discussed converting the largest of these barns into a home for their retirement.

In the past, Vic had extended a modest two bedroom bungalow to create a five bedroom home with a swimming pool, tackling all of the work himself over a number of years. His job designing and building golf courses means that he has an eye for detail and practical skills which he felt would enable him to undertake the majority of building work involved in converting the listed barn.

The Waglands sold the remainder of their barns to a builder in order to fund the project, retaining six acres on which they have garages and stabling for their horses. Their large oak framed barn had been used as a cowshed and was in a poor state of repair, with a makeshift tin roof and black stained weatherboarding. Inside, however, the impressive oak frame was largely intact, and the couple were determined to retain and preserve as many of these timbers as possible.

They contacted local architect John McCall, who had been involved in renovating the farmhouse, and asked him to help them design a layout for their new home. John had undertaken his own barn conversion, and his first-hand knowledge of working with old farm buildings proved invaluable.

Sue's daughters, Jessie, 15, and Toni, 19, both live at home, so the Waglands wanted to include five bedrooms on the first floor. "The planners asked that we leave a minimum of 25 per cent of the barn open to the roof, to enable views of the vaulted ceilings," Vic explains. "The sheer size of the building meant that we could fulfil this request by creating a full-height gallery in the centre of the house above the entrance hall and dining area, and still have enough space for five generous

"WE WERE DETERMINED NOT TO CUT CORNERS, OUR BIGGEST PRIORITY WAS TO DO THE BUILDING JUSTICE – REGARDLESS OF THE COST.

Oak flooring has been laid in the main entrance hall, where the staircase rises up to a bridge-like landing overlooking the dining room. ➤

Exposed brick walls, an Aga and Pippy oak units were chosen for the spacious farmhouse kitchen.

bedrooms and three bathrooms upstairs."

The ground floor living space is relatively open plan, with a large entrance hall leading directly into the dining room and a cosy snug area positioned behind the stairs. The brick chimney breast partially divides the sitting and dining rooms and, to the other side of the stairs, there is a spacious kitchen/breakfast room, a utility, shower, study and a new games room which replaces a ramshackle unlisted goat shed – forming a single storey wing at right angles to the main building.

Externally, the Waglands planned to introduce a minimal number of new window openings, cladding the structure with black rebated feather-edged boarding and roofing it with red handmade plain clay tiles to match other traditional buildings in their Hertfordshire village. Additional rooflights were specified for the master and guest bedrooms, which cast light through internal windows into the hallway below. They submitted their planning application which, following some debate regarding the number of windows, eventually received approval.

Vic employed a colleague who had worked with him for several years on the golf courses, and together the two men completed over 80% of the project themselves – bringing in specialist subcontractors such as electricians and plumbers when necessary. Work began the next spring and Vic dug service trenches for water and electricity using his own machinery.

"The barn was leaning so severely that we had to build a buttress to hold it up before we could even start work," says Sue. "When Vic began underpinning the building he discovered some old coins hidden inside a brick and buried in the footings. The dates coincided with the date on the beam, so we assume that the original structure was built in 1812. We took these coins, adding some millennium coins to bring its history up to date, and hid them in the new brick fireplace together with a note about the conversion."

An engineer specified elements of steelwork to secure the frame and roof structure, with stainless steel cross wires tensioning the timbers. The ends of most of the roof rafters were rotten but, rather than replace these, Vic removed the ends of the timbers and lowered the roof by a few centimetres in order to retain the original beams.

The Waglands wanted to keep as many of the old oak timbers as possible, cleaning the frame by hand over a period of six weeks. They retained all of the main framing members and – where new window openings were formed – a minimal number of studs were removed and ➤

"WE THOUGHT EVERYTHING THROUGH VERY CAREFULLY, AND SPENT A GREAT DEAL OF TIME LANDSCAPING THE GROUNDS TO MAKE A PLEASANT SETTING."

reused elsewhere in the building. Any additional studs have been replaced with oak, using traditional methods, with the majority of new timbers required on the west side where the structure was leaking badly. The walls and roof were then insulated externally so that the existing framing could be viewed in its entirety.

"We were determined not to cut corners," remarks Vic. "Our biggest priority was to do the building justice –regardless of the cost. This meant that we did run over budget, and it took the best part of two years to complete, but virtually everything is oak – from the internal doors to the panelling and skirtings."

"We stayed living in the farmhouse 50 metres from the barn throughout the build, and then sold it once we were able to move in," says Sue. "The fact that Vic completed so much of the work himself has made it even more special. His son is a master bricklayer, and built the internal walls and the fireplace in our living room. It required one-metre-deep footings, steel supporting rods and took three men to lift the oak bressumer beam into place, but the end result is stunning and it is an absolute certainty that it will never fall down."

The bespoke handmade kitchen was crafted from Pippy oak, which has subtle shakes and weatherworn clusters of knots or 'pips'. Andersons of Hitchin designed and supplied the units, including a granite-topped island which doubles as a breakfast bar.

"We saw four other kitchen designers, but Andersons impressed us by being the only company that suggested cutting the granite worktops to fit around the existing oak studwork," says Vic. "Everyone else wanted to create a plinth which would have concealed the wood."

Attention to detail is evident throughout the house. Vic bought 12 radiators salvaged from a London hospital for £60 each, which were pressure tested, sandblasted and sprayed with a powder-coated finish to match the bespoke iron light fittings. Such large, high spaces demanded grand lighting, and there are 57 bulbs in the lounge and dining room alone, which require a ladder to change them.

"We thought everything through very carefully, and spent a great deal of time landscaping the grounds to make a pleasant setting," says Vic, who dug a large pond, and fashioned lawns, steps and informal planting in what was previously a muddy field. "It was similar to shaping a golf course, and the contours of the garden served to hide the spoil from the site. I find it very satisfying being able to look out of the window and survey my work now that it's all over. It was a challenging project but one which we all thoroughly enjoyed." ■

FACT FILE

Names: Vic and Sue Wagland
Professions: Retired
Area: Hertfordshire
House type: Five bedroom converted barn
House size: 400m^2
Build route: Selves and subcontractors
Construction: Oak frame, clay roof tiles
Warranty: Zurich
Finance: Private
Build time: March '00 – March '02
Land cost: Already owned – value £300,000
Build cost: £375,000
Total cost: £675,000
House value: £1,500,000
Cost/m^2: £938

55%
COST SAVING

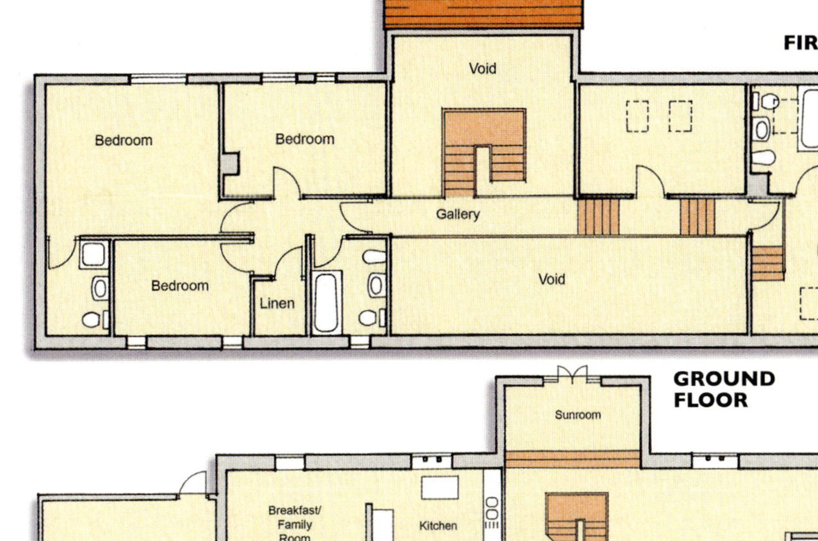

"IT WAS A CHALLENGING PROJECT BUT ONE WHICH WE ALL THOROUGHLY ENJOYED."

FLOORPLAN

A galleried reception hall opens into the snug beneath the stairs and a dining and living area, divided by the chimney breast. The kitchen/ breakfast room has a separate utility, and there is also a study, shower room and games room on this level, with five bedrooms and two bathrooms upstairs.

USEFUL CONTACTS

Architect John McCall: 01462 459213 **Stairs, windows and external doors** A&J Joinery: 01462 682888 **Kitchen design and installation** Andersons of Hitchin: 01462 437343 **Internal oak doors** Justin Wand: 01438 871775 **Roofing** Vessey & Johnson Roofing Services: 01438 226944 **Oak supply** Ternex Sawmills: 01707 324606 **Roof tiles** Tudor Roof Tiles: 01797 320202 **Building materials** Chas Lowe & Sons Ltd: 01438 812740 **Exterior timber, joists, materials** Jewsons: 01727 854588 **Light fittings Smithbrook** Iron Lighting: 01483 272744

LISTED BUILDINGS

Listed buildings are protected by law in their entirety, including all buildings within the curtilage that were built before 1 July 1948. It is a criminal offence, punishable by fines and possible imprisonment, to carry out any works of alteration or extension to a listed building in any manner which would affect its special interest, internally or externally. Even minor extensions and alterations that do not require planning permission require listed building consent. Local authorities have powers of enforcement to reverse or mitigate unauthorised works at the owner's expense. As some compensation for these restrictions, home improvement grants may be available from the local authority, and there are VAT concessions (zero-rating) on certain types of work classified as 'approved material alterations'.

A CONTEMPORARY STYLE CONVERSION OF AN OLD BARN

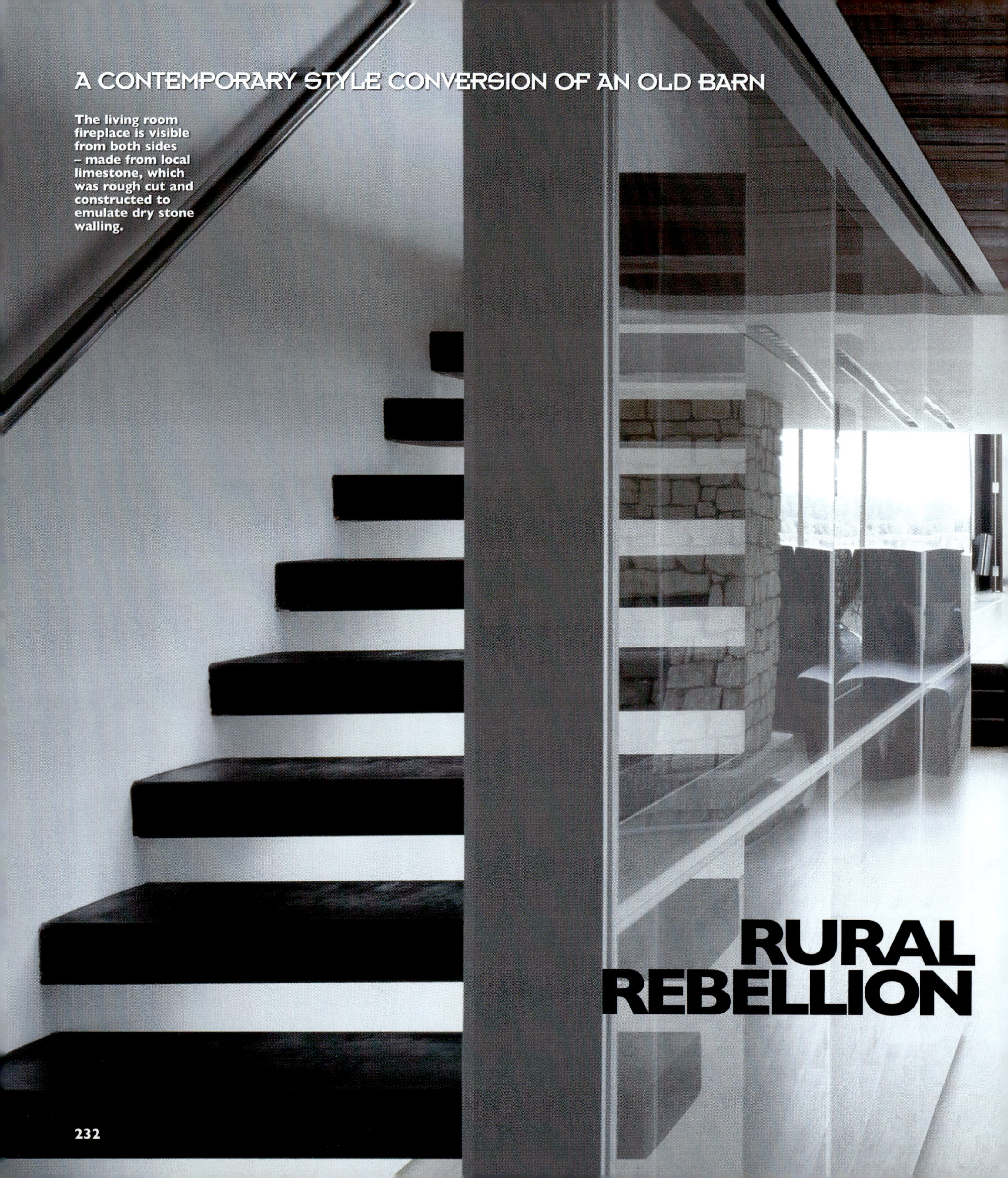

The living room fireplace is visible from both sides – made from local limestone, which was rough cut and constructed to emulate dry stone walling.

RURAL REBELLION

Nicolas and Alison Tye's project breaks the rules of barn conversions – and provides them with affordable contemporary interior space thanks to the remarkably innovative use of materials...

NICOLAS TYE AND his partner, Alison, had previously converted a former shoe factory in London's Clerkenwell, but hoped to move to a more rural location. "Although our apartment had plenty of space, there was no garden, and we're both originally from the country. We looked in different areas, but everything was too expensive. It was only when some friends suggested we try Bedford that we eventually found the barn," Nicolas, an architect, explains.

Dating from 1860, the structure had been used as a grain store to the local farm for the majority of its life before falling into dereliction. Set back from a country road along a private track, one mile from the nearest village, it sits on the crest of one of the highest points in Bedfordshire – enjoying panoramic views of the surrounding countryside.

Measuring 40 metres in length but just six metres wide, the oak framed building was set on a brick plinth and infilled with red bricks and timber cladding. A series of elm trusses had been used for the roof, with thick boarding running the full length of the barn.

"WE WANTED TO SLIDE A NEW HOME INTO THE OLD BUILDING, WHILE KEEPING ALL THAT WAS POSSIBLE OF THE EXISTING FRAMEWORK."

Pointers to the past **The metal funnel chimney, entrance gateway and canopy are designed as reminders of the barn's former agricultural use.**

"It was a great big cathedral-like space with fabulous views – a really winning combination," says Nicolas. "We wanted to slide a new home into the old building, while keeping all that was possible of the existing framework. It meant sacrificing two of the six potential bedrooms in order to achieve this, but the new layout opens up the ground floor living areas into one large, open plan space."

The barn originally covered some 275m^2 which, after excavating the ground floor to an almost semi-basement level and introducing a first floor, has increased to 387m^2. "The excavation meant we were left with a building 21/2 storeys high," Nicolas explains, "and in certain parts of the ground floor there are openings which give views right up to the roof's apex."

"Basically the entire building was stripped back to its frame and rebuilt with new brick and block cavity walls," says Nicolas. "The roof structure was sandblasted and sprayed with a flame retardant, and a new roof of insulation, counter battens, membrane and natural slates has been constructed like a hat over the whole thing."

A fully glazed gable forms the end point of the ground floor space, with dramatic views of rolling countryside and panoramic sunsets.

Affordable luxury: the 5 metre long kitchen unit was constructed from plywood with a stainless steel worktop.

"THE USE OF NATURAL, ORGANIC MATERIALS WHICH ARE HEALTHY TO HUMANS WAS UPPERMOST IN MY MIND WHEN I WAS DESIGNING THE CONVERSION."

Initially the couple employed a main contractor, who went bankrupt while the project was still at foundation stage. This prompted Nicolas to leave his job in London and work on site full time, project managing and labouring prior to starting up his own practice in a studio within the barn. "What started out as a catastrophe had a silver lining because I was able to learn so much about the construction process, and finally realise my ambition to start my own practice," he says.

The budget was tight, but Alison and Nicolas wanted to incorporate centralised control systems for the lighting, underfloor heating, vacuum, ventilation, heat recovery and computer systems built into the fabric of the barn. Ethernet and Cat5E cabling has been installed throughout the house, which allows CCTV, TV, satellite and music to be connected to all rooms, and there is a sophisticated alarm system with electronic entry gates.

In order to afford this modern technology they chose some inexpensive, extremely low tech materials which reflect the agricultural nature of the building – including staircases made from rusting metal, plywood kitchen units, partition walls built using £10 flush faced door blanks ➤

Cantilevered rusty metal stairs appear to float behind a double wall of Perspex set in an aluminium frame.

"LIVING SO HIGH UP MEANS THAT WE ENJOY FANTASTIC VIEWS AND PANORAMIC SUNSETS. IT REALLY IS A WONDERFUL SETTING."

and window surrounds constructed from shuttering plywood.

"The use of natural, organic materials which are healthy to humans was uppermost in my mind when I was designing the conversion, and I investigated a variety of non-toxic, low allergen, healthy products," says Nicolas, who chose organic paints and non-toxic stains. "We have used timber extensively throughout the project – after all, this is a timber framed building, with a golden syrup coloured timber boarded ceiling. The flooring is ash, and all the windows are Scandinavian sustainable timber with triple glazed units."

The ground floor is open plan, with kitchen, dining and living spaces flowing into one another. A laundry, utility and WC have been set against the back wall and benefit from high level light through large, sandblasted glass panels, designed to avoid overlooking the original farmhouse which is located to the rear of the property.

A two metre wide stone chimney stack, with fireplaces to both the ground and first floors, acts as a partition in the living areas and was built from rough cut local limestone designed to emulate dry stone walling. This is flanked by a staircase of cantilevered rusty steel treads, which appear to float behind an aluminium-framed wall of transparent Perspex.

On the first floor a bridge connects the four bedrooms and two bathrooms. "Each bedroom has its own views of open landscape and the lake, and the windows are framed to draw your eye out to the view, with recessed roller blinds and hidden lighting designed to flood the roof line above," Nicolas states.

Instead of sinks, water from taps set into the wall splashes onto a slab of stone, and each wet area has a white ceramic WC with an electronic flush system, beige ceramic floor tiles heated from the underside, and backlit mirrors offset from the wall.

Approached from the west, a fully glazed gable end overlooks the rolling countryside below, and may be seen from five miles away.

"The building acts as a beacon on top of the hill," says Nicholas. "But the shutters and blinds mean that we never feel exposed. Living so high up means that we enjoy fantastic views and panoramic sunsets. It really is a wonderful setting." ■

FACT FILE

Costs as of Dec 2004
Names: Nicolas and Alison Tye
Professions: Architect and doctor
Area: Bedfordshire
House type: Four bed converted barn
House size: 387m²
Build route: Self and subcontractors
Construction: Oak frame, masonry walls, slate roof
Warranty: Architect's Certificate
finance: Chesham BS mortgage
build time: July '02 – March '04
land cost: £245,000
build cost: £242,000
total cost: £487,000
house value: £1,000,000
cost/m²: £625

51%
COST SAVING

FLOORPLAN

The barn has been increased by 100m², with four bedrooms and two bathrooms on the new first floor level and open plan living areas on the ground floor, some of which have been left open to the roof.

GROUND FLOOR

FIRST FLOOR

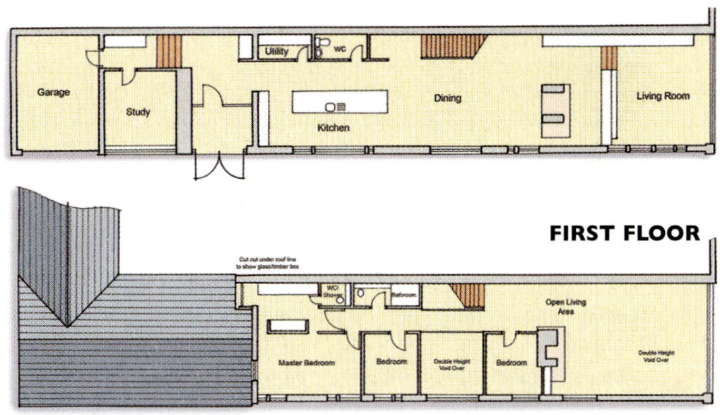

USEFUL CONTACTS

Lighting – Modular: 0207 6819933; **Metalwork** – Dennis Hawkins Metalwork: 01908 551400; **Flooring** – Tarkett Timber Flooring: 01905 342700; **Flooring** – GSM Stone: 01580 212222; **Kitchen worktops** – GEC Anderson: 01442 826999; **Vacuum system** – Beam: 01905 611041; **Blinds** – Soltech Systems: 01628 776488; **Perspex and aluminium wall** – At Work Associates: 0207 6083737; **Home cabling** – First Quantum: 0208 7498486; **Windows** – Tanums Fonster Swedish Windows: 0115 9321013; **Sliding timber doors** – I-D Systems: 01603 408804; **Furniture** – Atrium: 0207 3797288; **Ironmongery** – Allgood: 08706 090009; **Building supplies** – Jewsons: 01525 713811; **Air systems** – Villavent: 01993 772270; **Lighting control** – Lutron: 0207 7020657

AN ELEGANT SOLUTION

Philip and Angela Traill have created an innovative contemporary home within the shell of a 19th century timber framed barn.

WORDS: MICHAEL HOLMES PHOTOGRAPHY: PHILIP BIER

WHEN ANGELA TRAILL'S husband Philip told her that she would never be able to fit both the free-standing bath she wanted in their new en suite bathroom as well as a large walk-in shower enclosure, the disappointment only made her more determined to prove him wrong. At ten o'clock one night, she cut scale templates for each item from newspaper, went over to the barn that they were converting into their new family home, and with only the light of a small bedside lamp borrowed from one of her children's bedrooms, marked out the exact positions for the shower, bath and vanity unit on the bathroom floor.

"It turned out that there was more than enough space in there for everything," recalls Angela. "In fact, if anything the room has turned out to be extremely spacious – but I just had to make sure that it would work."

Angela's actions that night were typical of the degree of planning and forethought that she and Philip, who have two young daughters, applied to the whole of their project, the conversion of a 19th century Victorian threshing barn on the edge of a village in rural Surrey.

Philip admits to being a perfectionist who kept a tight reign on every aspect of the design and build. There was however an added incentive to complete the project on time – theirs was one of eight self-builds followed by the new series of Channel 4's *Grand Designs*. Philip and Angela had ➤

"THE BEDROOMS ARE AT EITHER END OF THE BUILDING, LIKE GALLERIES, LEAVING THE CENTRAL SPACE ESSENTIALLY UNALTERED."

committed to completing their home within just six months to suit the programme's strict filming schedule. Given the enormous complexity of the design, this was a hugely ambitious undertaking, but Philip and Angela were determined not to fail.

"The project was tremendously challenging, but also very exciting," recalls Philip. "There were days when the television cameras were a real handicap and work came to a virtual standstill. But on the whole, TV helped to make things happen on time.

"Building regulations decisions, for instance, were made by the top man and we got very comprehensive answers. It also helped that the majority of suppliers focused on timely and complete delivery – an enormous help in any self-build project."

Whilst they must take a great deal of the credit for the project's success, Philip and Angela are very keen to point out the critical roles played by their architects and their main contractor, Peter Shoesmith. "We were very lucky to pull together such a great team," says Philip. "People tend to come to a project with different skill sets and I think we managed to find just the right balance."

Concept design architect Damien Blower's design for the Traills' barn was particularly inspired, making full use of the timber framed structure's height to create a home over three storeys where most designers would have settled for two. It is a bold, exciting and original treatment of the space, made all the more interesting by the use of sweeping curved walls within its interior.

The great challenge posed by a barn for conversion is where to position the first floor accommodation without compromising the sense of space and volume that is invariably the building's greatest asset. With a threshing barn there are also the two great cart door openings to contend with. The principle of Damien Blowers' solution was straightforward enough, to position the bedrooms in two sections at either end of the building, like galleries, leaving the central space essentially unaltered. Instead of creating conventional timber and plasterboard boxes, however, his design houses the bedrooms within elegantly curved structures that assume an almost sculptural form and which the Traills refer to as 'pods'. To overcome the ➤

To overcome the impracticality of having two independent sections of bedroom accommodation, they are linked by a bridge landing that sails over the main living area.

The second floor master bedroom is housed above its en suite, within the smaller of the two bedroom 'pods'. It enjoys spectacular views.

"THE PLANNERS WANTED THE CONVERSION TO REMAIN INVISIBLE FROM THE ROAD AND FROM THE CHURCH ACROSS THE FIELDS."

impracticality of having two independent sections of bedroom accommodation, inappropriate for a four bedroom family home, there is a bridge landing that sails over the main living area.

Access to the first floor is via a geometrically perfect helical staircase, made from steel, oak and plaster, that rises from the centre of the living area and emerges at the junction of the bridge and the first floor landing. Here the larger of the two 'pods' houses the two family bedrooms and bathroom, with a further small staircase leading up into the roofspace above and the guest bedroom. There is also an enclosed staircase down to the utility area below, necessary to satisfy the fire regulations for a three storey building – an addition made by Elspeth Beard in the building regulations phase. At the opposite side of the bridge, in the smaller 'pod', is the master en suite and above that the master bedroom.

In contrast with the vast proportions of the main central living area, the two spaces below the bedroom 'pods' form more conventionally proportioned living spaces. The smallest has been enclosed to create a separate cosy living room where the family can watch television or listen to music. This room also features a double fronted woodburning stove which it shares with the main living area.

The corresponding space on the opposite side of the barn is open to the main living space, and is used as a formal dining area. This space also forms the link and transition into the farmhouse style kitchen/breakfast room, housed in a newly built single storey annexe.

"If we had owned only the threshing barn, the kitchen would have to have been sited within the main living area, but we were fortunate enough to own a small adjacent stable block," explains Angela. "Because it lacked foundations, this had to be entirely rebuilt but it was perfect for housing the kitchen and breakfast room. It was the historic and listed buildings officer who proposed the solution of adjoining the two buildings using a slightly recessed glazed link to ensure that they appeared to remain separate."

Getting the planners to agree to allow the two buildings to be linked was a relatively minor achievement compared to gaining the initial consent for the principle to convert the barn into a dwelling. ➤

The kitchen, designed and built by master craftsman Lionel Daniels, is built around a traditional Aga range and features curved units built in Sycamore with a slate top from Kirkstone Quarry. The central panel of every cupboard features an individually selected piece of timber veneer, each one from a different species.

"WORKING ON AN OPEN CONTRACT MEANT THAT WE HAD NO GUARANTEE WHAT THE FINAL COST FOR LABOUR WAS GOING TO BE."

The bridge landing is formed using rolled steel and rests on new blockwork walls. The couple wanted to avoid any columns in order to retain the central space uninterrupted.

"First of all we had to establish the justification for conversion to residential use," explains Philip. "Local plan policy favours the conversion of redundant farm buildings for commercial use and so we first had to prove that the building was unsuitable for this purpose due to restricted space for parking.

"We used a planning consultant to argue this for us and only succeeded after presenting the council with a case law precedent. From then on it was a matter of ensuring that the barn retained its character. The application was made all the more sensitive by the fact that the barn is within the curtilage of a grade II listed building, on green belt land, in an Area of Outstanding Natural Beauty and is also a site of archaeological interest – we had to commission a watching brief at a cost to us of around £5,000. The planners wanted the conversion to remain invisible from the road and from the church across the fields. We had to justify the need for every single window opening. In all the planning process took seven to eight months."

Securing building regulations approval for the ambitious design was to prove equally challenging. At this stage the Traills retendered the work and engaged architect Elspeth Beard to take over the building regulations applications process and construction phase. "Elspeth turned out to be an excellent choice, as she had already experienced many of the same problems that we faced during her own project, the conversion of a disused water tower," says Philip. "She was extremely creative and understands the level of detail that builders require."

The Traills' contractor, Peter Shoesmith, was one of four builders asked to tender for the work and was hired on an open contract under which he committed to deliver the project to a set schedule for a fixed fee, with a substantial bonus for meeting the completion deadline.

"Rather than building a profit into his costs, Peter agreed to work for a fixed monthly fee and to declare all of the labour and material costs openly," explains Philip. "This way of working gave us the flexibility to make decisions as we went along on materials and finishes. It enabled us to bring in our own nominated subcontractors for certain work, such as the kitchen, and to get involved on a DIY basis. Putting all of the costs through a VAT registered contractor also meant we were also able to get all of the work zero rated at source, rather than being charged five per cent and having to wait until completion to reclaim this. This helped with cashflow.

"Unfortunately, working on an open contract also meant that we had no guarantee what the final cost for labour was going to be and this did eventually exceed our expectations, putting pressure on our budget. That was certainly one of the few low points in the project."

To help bring costs back in line, Philip and Angela got involved in some of the building work themselves, including plasterboarding and non-volt wiring (Cat 5e). Philip designed some of the lighting himself, using a variety of low voltage spotlights and, in the main living area, a

Contemporary style basins and bath are from bathstore.com, with taps from Vola. Angela desgned the table and timber bath surround. The curved 'snail' walk-in shower enclosure is from WEDI systems and the tiles from World's End Tiles. ➤

"THE BEST PART IS WAKING UP AT NIGHT, CROSSING THE WALKWAY AND SEEING ALL OF THE SPACE AND THINKING: THIS IS TRULY ALL OURS."

'chameleon light' which can be programmed to gradually change colour. Philip and Angela also installed miles of cabling for a number of home automation features, all of which terminate in the utility room, the nerve centre of the house. Here are amassed the circuits, switches and fuse boxes for all of the other gadgetry including programmable thermostats for each of the twelve underfloor heating zones, the transformers and switching units for the automated lighting, the patchbox for the data network, the incoming ISDN lines, TV signal amplifier, CD player for the multi-room hi-fi, the security system and more. Philip has even installed a link which allows him to control the heating and lighting from anywhere via the internet.

Although they have a passion for technology, the Traills are also committed to an organic and chemical-free lifestyle, and consequently all of the paints and stains used in the project were organic, as are the carpets. They also decided to limit their use of processed water by installing a rainwater harvesting system, a vast tank buried in the garden next to the garage, a converted cart shed.

The day before the television cameras returned for their final shoot, seven months after work had started on site, was one of the most frenetic. "It still looked like a building site," recalls Angela. "But we pulled out all of the stops to be ready in time. Turf was laid outside, the decorating was finished, everywhere tidied and the sofa finally taken out of its wrapping. By the time the cameras rolled, it looked perfect – even though there is work that still needs completing to this day.

"We really enjoyed being involved in the programme, and the entire project," says Angela. "It was extremely hard work but it is just fantastic living here. The best part is waking up at night, crossing the walkway and seeing all of the space and thinking: this is truly all ours." ■

FACT FILE

Costs as of May 2003
Name: Philip and Angela Traill
Profession: Marketing Strategist and Freelance PR Consultant/Housewife
Area: Rural village in Surrey
House type: Converted Victorian Barn (Grade II listed)
House size: 280m^2 + Office & Garage
Build route: Main Contractor & DIY
Construction: Timber and brick
Warranty: Zurich Conversion I
Sap rating: 103
Finance: Bank
Build time: Jun–Nov '02
Land cost: Already owned
Build cost: £450,000
Cost/m^2: £1,607

Cost Breakdown:

Total Project Cost	£450,000
Including; Kitchen	£30,000
Bathrooms + Shower room	£17,000
Lights	£15,000
Windows, sliding barn door and glass roof	£34,000
Helical Stair & handrails	£15,000
UFH and Ventilation (inc. fitting)	£15,500
Rain Harvesting System (Purchase only)	£3,900
Double fronted fireplace	£2,400
Archaeological Watching Brief	£5,000
Total	**£450,000**

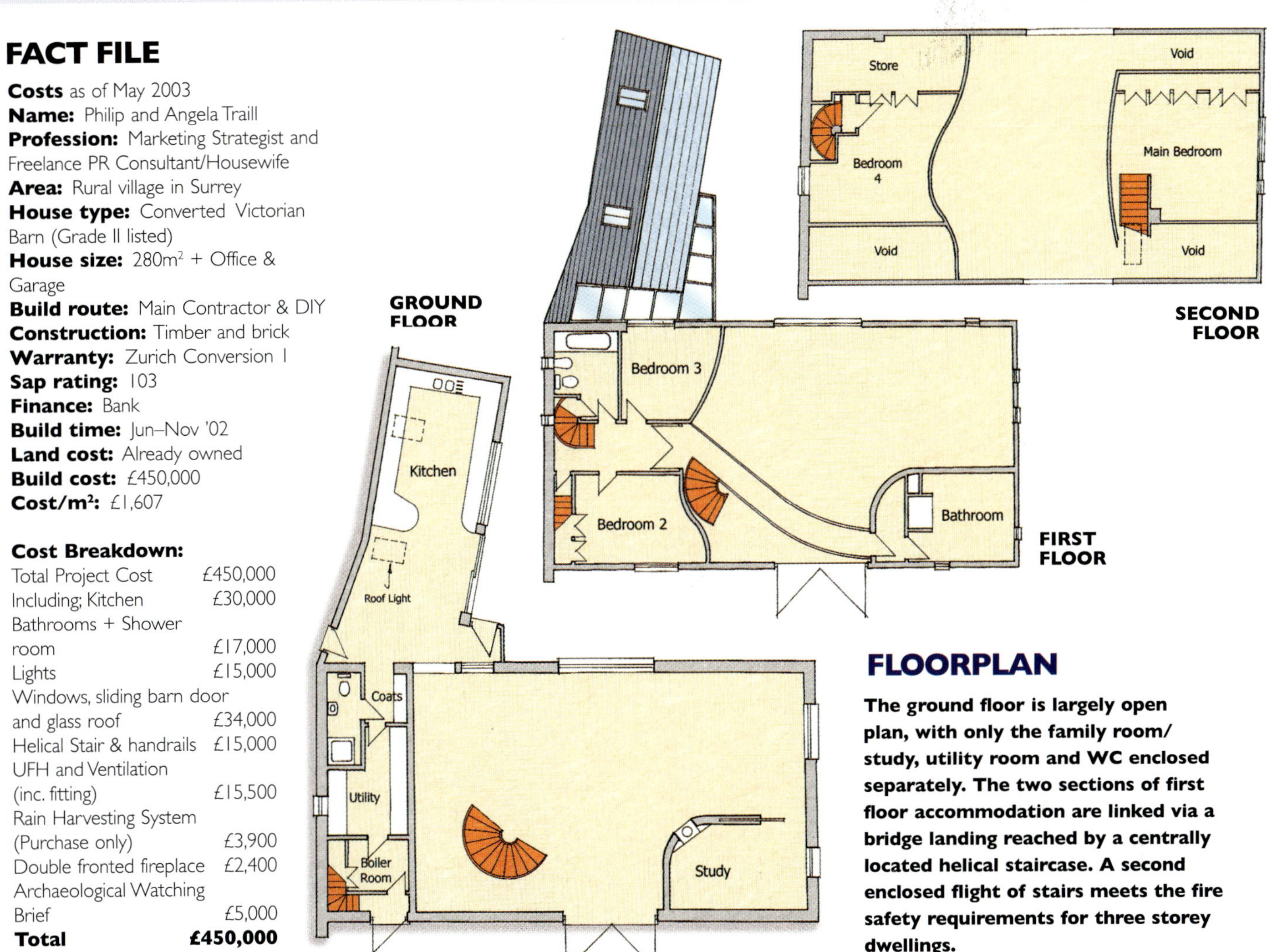

FLOORPLAN

The ground floor is largely open plan, with only the family room/ study, utility room and WC enclosed separately. The two sections of first floor accommodation are linked via a bridge landing reached by a centrally located helical staircase. A second enclosed flight of stairs meets the fire safety requirements for three storey dwellings.

USEFUL CONTACTS

Concept Design Architect - Steadman Blower: 01252 783574; **Detailed Design and Project Architect** - Elspeth Beard: 01483 860342; **Structural Engineer** - W.G. Read & Assoc.: 01483 537370; **Building Contractor** - Peter Shoesmith: 01483 210250; **Helical Stairs (Design, supply & install)** - KDA: 0208 806 8399; **Bespoke Steel Windows** - West Leigh Ltd: 020 7232 0030; **Bespoke Sliding Glazed Window (for Barn Door)** - Vista-Brunswick Ltd: 0117 955 1491; **Steelwork - Substructure & Walkway** - A J Fabrications: 01483 276016; **Conservation Rooflights** - The Metal Window Co.: 01993 830613; **Garage Door Frame** - Hormann (UK) Ltd: 01530 513000; **Kitchen Units** - Daniels & Eldridge: 01730 827472; **Kitchen Slate Worksurface** - Kirkstone Quarries Ltd: 01539 433296; **Range Cooker** - Aga: 01952 642000; **Stone Flooring** - Mastacraft: 01420 560060; **Sanitary Ware** - bathstore.com; 07000 228 478; **Spiral Shower Enclosure** - Wedi: 0870 9907504; **Taps and Showers** - Vola UK Ltd: 01525 841155; **Heated Towel Rails** - Myson Towel Wamers: 01204 863200; **Bathroom Floor and Wall Tiles -** World's End Tiles: 0207 819 2115; **Glass Double Fronted Stove** - Anglia Fireplaces: 01223 234713; **Oak Flooring** - Joachim Eckert Wood Floors (UK) Ltd: 01252 520520; **Stainless Steel Support Rods (Walkway)** - Spencer Rigging Ltd: 01983 292022; **Insulation** - Kingspan: 0800 610061; **Organic Builders Merchant** - Construction Resources: 0207 450 2211; **Architectural Lighting** - DeltaLight (UK) Ltd: 01428 651919; **Lighting Control Solutions** - iLight Ltd: 01892 870072; **Underfloor Heating** - Nu Heat: 01404 549 770; **Temperature Network Control System** - SmartKontrols Ltd: 01825 769812; **Mechanical Ventilation** - SpaceAir Solutions Ltd: 01483 504 883; **Condensing Boiler** - AquaFlame: 01953 454896; **Home Network** - RW Data: 01604 706633; **Ironmongery** - Joseph Giles: 0208 680 2602; **Damp-Proofing & Timber Preservation** - Terminix Ltd: 0800 789500

CONVERTING A LISTED BARN IN CONTEMPORARY STYLE

BARN BEAUTIFUL

Julian and Ali Pilling have created an elegant and spacious contemporary style home out of a derelict barn.

WORDS: HEATHER DIXON PHOTOGRAPHY: DAVE BURTON

THE FIRST THING visitors comment on when they visit Julian and Ali Pilling's converted barn in North Yorkshire is not the spectacular views across rolling countryside or the peaceful location of their converted barn – but the staircase.

As one of the founding partners of Helmsley-based engineering designers Bisca, who specialise in canopies and staircases, the last thing Julian wanted was conventional access to the upper floor. So he created a £15,000 focal point in the open plan conversion with green English oak, textured steel ➤

"I WAS WORKING EVERY HOUR OF THE DAY... BUT IT WAS WORTH THE EFFORT."

uprights and cantilevered treads secured into a 2' thick stone wall.

The staircase is one of the key features in a property which has been renovated to emphasise, rather than disguise, its natural attributes – including untreated internal stone walls, reclaimed maple wood flooring and original beams which zig-zag into the vast roof spaces.

It has taken Julian the best part of 10 years to renovate the barn in a triple-phase project which began with the renovation of a farmhouse, now owned by members of his family, followed by the development of a two-bedroom cottage and workshop from a neighbouring dilapidated farm building which had stood derelict for 25 years.

Although the 200-year-old property is a listed building, Julian initially applied to Ryedale District Council for planning permission to build an annex at the back of the property, where he could live, while using the rest of the building for the development of his then fledgling business with partner Richard Mclane.

"It was a lot more difficult to get planning permission than I thought," says Julian. "The council was opposed to new housing and this involved a new extension to the main barn, so I employed a planning consultant and we pushed very hard to get the application passed."

The difficulties didn't end there. When Julian wanted to move Bisca out of the farm buildings into nearby Helmsley, and convert the workshop into a house, he had to prove to the planning department that the farm building was redundant and could no longer be used for commercial purposes.

"Having just run our business from there, it could have been very tricky but, luckily for us, the planners had received a number of complaints about the workshop being located in the barn so they were in a Catch 22 situation."

Julian also had to convince planners that he wanted to work with the character and materials of the existing building, not create a carbuncle on the beautiful North Yorkshire landscape. "We

Julian and Ali designed a niche to house their wide screen television into the chimney breast. ➤

"WE TRIED TO USE LOCAL CRAFTSMEN AS MUCH AS POSSIBLE, AND PEOPLE WE KNEW. IT MEANT THE WHOLE THING WAS PROBABLY DELAYED BY ABOUT SIX MONTHS AS A RESULT..."

The cantilevered Bisca staircase is made of English green oak. The garden room beyond was originally a garage.

wanted a family home with five bedrooms and a large kitchen, but with an open plan feel," says Julian. "The building had to be taken back to basics."

This included reroofing with the existing pantiles, which were individually cleaned and reversed, partly replacing the roof timbers, which were cleaned with a stiff brush and sealed with matt varnish, and installing concrete floors. All the exterior and interior walls were repointed and many of the key walls left as bare stone inside to add to the rustic element of the conversion. Other walls were plastered to develop textural contrast and key walls painted in bold colours to create striking divisions between the rooms.

"We decided where the key rooms were going to be and built everything up around them," says Julian, who drew the plans himself and employed a local architect to create the final drawings. "We chose the single storey part of the building, which had once been a wheelhouse, as the kitchen area because it had views on three sides."

The master bedroom and sitting room were created by dividing the main barn with a steel-reinforced floor, leaving one part open from floor to ceiling for a double height dining section.

"The aim was to keep everything as open as possible, to take the eye naturally from one room to another and to retain all the character of the barn without losing its contemporary edge. Although the house is quite square and geometrical, there are steps and corners to keep it interesting and draw you through it."

The property has been well insulated and the Pillings managed to source reclaimed radiators which were granite blasted and powder coated. Indian flagstones in the hallway were treated with a silicone sealant to complement the texture and colours in the 25' high internal stone walls.

"A lot of people think bare stone inside is very harsh, but these are cosy rooms," says Ali. The property is heated from a 330 litre Mega Flow water heater and Firebrand boiler, stacked up in a cupboard to save space.

To save money and to make sure they achieved exactly what they wanted, the Pillings carried out as much of the labour as possible, oversaw ➤

Julian and Ali designed the kitchen without overhead cupboards and created a dividing wall to head height so they could see the mezzanine landing and open roof trusses.

Bisca handrails add form to the bridge which overlooks the dining area on one side and the hallway and kitchen on the other.

"THE BEAUTY OF GREEN OAK IS THAT IT CONTINUES TO SLIGHTLY CHANGE ITS SHAPE OVER THE YEARS SO YOU ALLOW FOR THAT NATURAL PROCESS IN THE BUILD."

the project between them and co-ordinated the team of locally-based joiners, electricians and roofers. They also sourced all the materials themselves, including six tonnes of maple floorboarding from an advertisement in the local paper. The Pillings simply laid it and sealed it.

"It's been tremendously rewarding," says Julian. "It took 18 months to achieve the second phase of the property and I was working every hour of the day to divide my time between the business and the renovation, but it was worth the effort. We lived without a kitchen for a year and had to get by with just a microwave and a kettle, but it's all about keeping your eye on the end result.

"We tried to use local craftsmen as much as possible, and people we knew. It meant the whole thing was probably delayed by about six months as a result, but we would rather do that than compromise. It's very easy to take on tradesmen just because they are available, but you need to deal with people who have been in the business a long time and who know what they are doing."

The staircase represents one of the most painstaking jobs in the house. Made of English green oak, each tread has been cut in slightly over-sized blocks and left to season. They were then cantilevered into a steel plate set in the stone wall. All the uprights were heated until they were 'red hot' and power hammered into shape, and the staircase finished with an elephant foot newel post and seasoned oak rail.

"The beauty of green oak is that it continues to slightly change its shape over the years so you allow for that natural process in the build," says Julian. "You end up with something unique and very tactile, which is why we wanted the stairs to be the focal point of the house. When you walk through the door they are the first thing you see. They work with the existing timbers and pull the whole living space together. Once the stairs were right, everything else seemed to fall into place." ■

FLOORPLAN

The living areas are as open plan as possible. The bedroom are arranged on both floors.

GROUND FLOOR

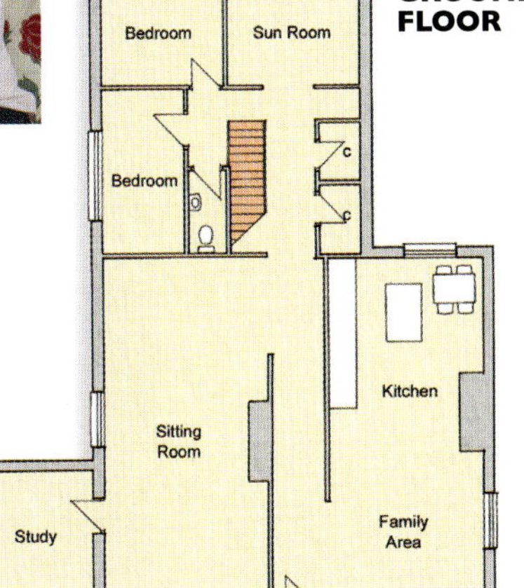

FACT FILE

Costs as of Dec 2003
Names: Julian and Ali Pilling
Professions: Director
Area: North Yorkshire
House type: Converted barn
House size: 325m²
Build route: Self-managed subcontractors
Warranty: NHBC
Construction: Stone walls with pantile roof
Finance: Private
Build time: 10 years in total
Barn cost: £50,000
Renovation cost: £160,000
Total cost: £210,000
Current value: £750,000
Cost/m²: £492

72 %
COST SAVING

BEFORE

FIRST FLOOR

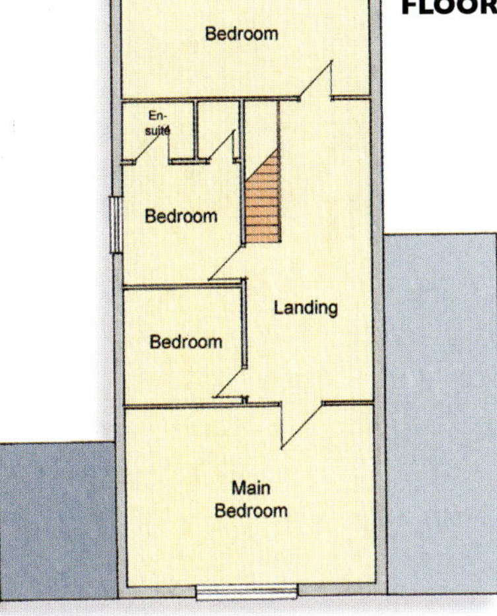

USEFUL CONTACTS

Granite worktops – F Jones:01642 241195; **Ironmongery, staircase and lights** – Bisca 01439 771702; **Rosedale Homes** - Rob Thomas 01751 472 050; **Bathroom** - Italian Bette bath from Travis Perkins: Shower is Hans Grohe:0870 770 1972; **Tiles** – Reed Harris, London 0207 736 7511; **Kitchen** – Phil Chandler from Carlton House Furniture:01845 526652; **Windows** – S.Taylor and Sons:01751 472143; **Stone Flags** – North West Reclamation 01282 603108

The barn is set in a village location and affords unbroken country views of the Lake District National Park, which inspire Alison and Jonathan in their work as landscape painters.

CUMBRIAN CONVERSION

Alison Critchlow and Jonathan Trotman have created a spacious yet low cost family home in the Lake District by converting an old barn.

WORDS: DEBBIE JEFFERY PHOTOGRAPHY: JEREMY PHILLIPS

MANY SELF-BUILDERS WOULD not think twice about spending £15,000 on a luxurious new kitchen but, for Alison Critchlow and Jonathan Trotman, this was their budget for converting and furnishing an entire barn. By tackling every aspect of the project themselves and re-using the existing materials, their new architect-designed home boasts handcrafted windows, slate floors and an open plan first floor living space which takes full advantage of the stunning Lake District location.

"We fell in love with the barn and immediately realised its potential," recalls Alison. "The views are panoramic, and we thought it would be a fantastic place to live." She and Jonathan, who are both landscape painters, had happened upon a photograph in an estate agent's window and decided to take a look. The slate structure was built in the 19th century as a traditional Cumbrian bank barn to house cattle, with a hay store above.

Detailed planning consent had already been obtained for a conversion, but it still took almost a year to buy the building, as the legal side of the purchase proved complicated. It was unclear who owned the land in front of the barn, over which Alison and Jonathan needed rights of access. Eventually the couple paid £35,000, some of which they borrowed from the Royal Bank of Scotland – the only lender prepared to oblige.

"We had never done anything like this before, but Jonathan and I both felt that the building was in such a stunning location that we should take the chance," says Alison. "My father is an architect, and offered his time and experience for free – otherwise we could not have contemplated such a project."

Rex Critchlow of Pye Critchlow Architects prepared the new plans, which were approved by the Lake District Planning Board and local building inspectorate. "We were keen not to alter the original building too much," says Alison. "My father treated us like clients, suggesting various alternatives and producing several sketches, and we chose the one which best suited our needs and involved the least intervention. It also ensured that we were granted planning consent quite easily as, from the outside, the barn remains little changed."

Internal walls have remained in their original positions, and the design incorporates one large living/dining/kitchen space on the upper floor with an open volume ceiling. One third of this area has been screened off for studio, library or guest use with storage above. The living space has a large window overlooking the distant hills, with a flush rooflight inserted into the southern slope. Other openings have been minimised and windows set back and stained black, retaining the barn's predominantly blank appearance. Two bedrooms, the bathroom and utility are located on the ground floor, where the entrance gives access to a new staircase in the south east corner.

Jon and Alison are both fine art graduates in their 20s, who calculated that their maximum budget for the conversion was just £15,000. In order to keep costs down they decided to undertake absolutely everything themselves, taking on plumbing, electrical work, damp coursing, drain laying and plastering as well as making the windows, doors and staircase. Their only labour expense amounted to just £100 for help with the plastering but, apart from this, they taught themselves all other trades by reading numerous books and magazines and attending the Homebuilding & Renovating Show. Rex Critchlow was also a valuable source of information regarding the structural requirements. "It was a very steep learning curve," admits Jonathan, who had previously taken a joinery course. "It was winter and, with no water or electricity on site, the first priority was to get connected to mains services. "Some very good friends filled our water bottles and let us use their bathroom," laughs Alison. "Jonathan and I dug the trench ourselves hiring a digger, which Jon taught himself to drive. It took him half a day just to get it going forwards! Whilst we were digging we hit an enormous pipe which drains the stream running beside the barn and, once this had been mended, it was necessary to dig beneath it. We then mucked out the barn and dug out the floor level downstairs to give us more headroom. Luckily my part-time job at a pottery didn't require an immaculate appearance!"

All materials were re-used and local recycled materials incorporated into the design, ensuring that the low-tech conversion is extremely eco friendly.

Cobblestones and cattle stalls were all carefully removed and saved, with block lining to below ground level and a concrete slab poured inside the building. The existing 600mm thick slate walls had been precisely laid so that each stone slopes slightly downwards in order to drain rainwater away from the building, and it was important not to interfere with this design. Once the limewash had been removed internally, a weak lime mortar was used to point the walls on the inside, with the external stonework left untouched.

"The roof was actually in very good condition," remarks Alison, "and we were able to simply clean the timbers and replace a few slipped and broken slates, using those taken from the new roof window aperture. Instead of

"OUR REWARD FOR ALL THE HARD WORK IS A UNIQUE PLACE IN WHICH TO BRING UP A YOUNG FAMILY IN INSPIRING AND BEAUTIFUL SURROUNDINGS."

stripping the roof and felting it, we decided to apply a traditional mixture of lime putty and cow hair to the back of the slates. It proved quite a messy operation, and we both ended up in A&E after getting lime into our eyes – despite wearing goggles." Two layers of rigid foam insulation, purchased cheaply as 'seconds', were then fixed between the rafters, which remain exposed in the first floor living area.

As the project neared completion, funds were running perilously low, and radical cost saving ideas were implemented. The kitchen cost just £170, as a neighbour had some old units stored in a shed. New door fronts were added and the old oak cattle dividers cleaned and used between the cupboards. Cobbles taken from inside the barn were laid to provide hard standing externally, with stone from the new window openings used to build a boundary wall. One of the old doors is now used as a panel for a cast iron bath, which was free to take away, while the toilet cost £15 and the bathroom sink £10.

"If we did need any additional materials we tried to find old items – such as the stone lintels for the bedroom windows which had once been kerb stones in a graveyard," Alison explains. "Our biggest challenge was positioning the huge stone lintel required for the upstairs window. It was so heavy that we needed nine friends to help us grapple it into place."

"The amount we learnt in two years is immeasurable," says Jonathan, who admits that the project dominated their lives. "Our reward for all the hard work is a unique place in which to bring up a young family in inspiring and beautiful surroundings. Plus we have an intimate knowledge of the building from the drains to the roof!"

Completed on time and on an amazingly low budget, the building has changed little in external appearance, the finishing touch being the re-liming of the front doorway surround. "Since the conversion was completed we have had two children," says Alison. "The barn is therefore a little cramped, but we have just acquired the small barn next door, which we hope to turn into a studio. Needless to say, we will be tackling all of the work ourselves!" ■

FACT FILE

Costs as of Nov 2003
Names: Alison Critchlow and Jonathan Trotman
Professions: Artists
Area: Cumbria
House type: Two bedroom barn conversion
House size: 74m²
Build route: Selves
Construction: Slate walls and roof
Finance: Royal Bank of Scotland
Build time: Two years
Barn cost: £35,000
Build cost: £15,000
Total cost: £50,000
House value: £195,000
Cost/m²: £203

74%
COST SAVING

Cost Breakdown:

Services	£1,850
Plumbing and drains	£1,100
Steel	£110
Floor Tiles	£380
Insulation	£800
Woodburning stove	£500
Flue	£460
Blockwork walls	£330
Reclaimed maple floors	£750
Electrics	£750
Glazing and roof windows	£900
Water heater	£600
Decorating	£400
Ironmongery	£200
Timber, including doors, windows and stairs	£1,600
General building materials	£2,600
Stone for cills and lintels	£150
Kitchen	£170
Bathroom	£200
Storage and panel heaters	£150
Equipment hire	£900
Labour	£100
TOTAL	**£15,000**

GROUND FLOOR

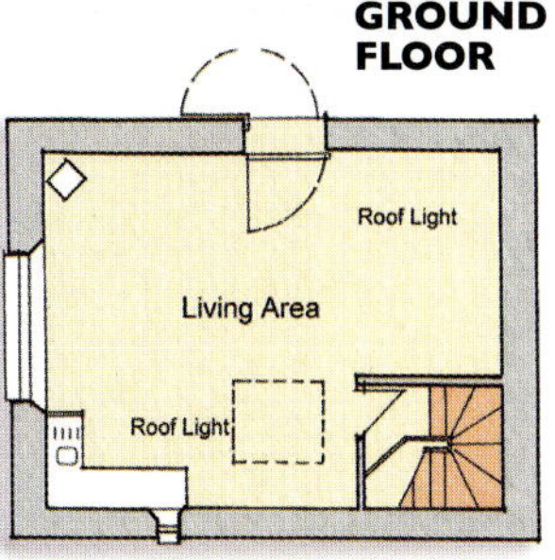

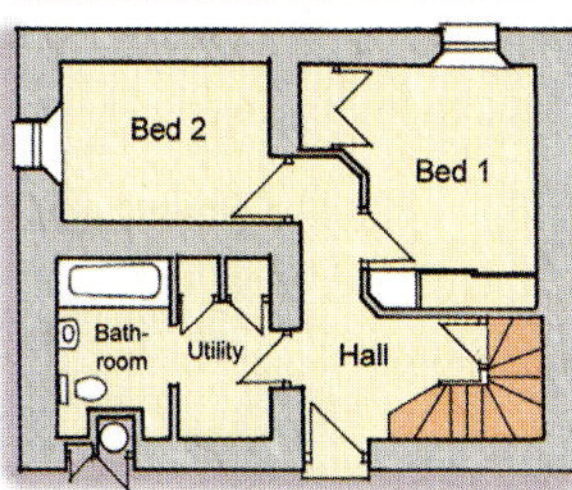

FIRST FLOOR

FLOORPLAN

The 19th century Cumbrian barn housed cattle below with a hayloft above. Now, two bedrooms and a bathroom are located on the ground floor, with an open plan kitchen and living room to the first floor taking full advantage of the stunning views.

USEFUL CONTACTS

Architect – Pye Critchlow Architects: 01472 350013; **Insulation** – Seconds & Co: 01580 767702; **Water heater** – Heatrae Sadia: 01603 420100; **Glazing** – Solaglass: 02476 547400; **Alison Critchlow and Jonathan Trotman: www.blencathraart.co.uk**

RENOVATING A CONVERSION

Gillian Ashby has remodelled an already converted but rather unloved barn to create her perfect home.

WORDS: VICTORIA JENKINS PHOTOGRAPHY: COLIN POOLE

GILLIAN ASHBY DID not intend to buy the converted barn that had just come onto the market in Devon. "I thought I wanted something small and easy to manage," she says. "But when I walked in and saw the 40 foot long dining room, open right up to the oak beams in the roof, and realised all that space could be mine, I couldn't resist it.

"It was wonderful – with big double barn doors and an enormous window looking out onto the courtyard. It was all so light, and had been extended to provide even more space." However, It was not in the best of conditions.

"It had been a barn for at least 140 years and had then been turned into a two storey residence by a builder," she explains. "For a short time it was a holiday let and then was bought and occupied by a lady who lived there for many years. By the time she left, it needed a lot of care and attention. There was a great deal of damp, so first of all I had to put in a damp-proof course. It also needed a bit of plastering and the layout needed rearranging too."

Fortunately Gill is very much

The position of the runner on the landing (right) shows where the builders linked the two galleries so that Gill could dispense with one of the two staircases. ➤

The dining room, into which the front door opens, has been reduced to 11m long after the kitchen was enlarged by moving a wall.

a hands-on homeowner, who saw the barn as a challenge so – where she could – she did the work herself. This included everything from mixing the concrete to tiling the walls, from decorating throughout to doing some of the plastering. "I've even done some of the plumbing," she says.

"The main problem was in the dining room, where there were two staircases – one at each end leading up to two galleried landings," she explains. "One led to the main bedroom, the other led to three more rooms plus a bathroom. You couldn't move from the main bedroom to the other three without coming down the stairs and up again."

Gill had the builder remove one of the staircases and extend the galleries until they joined up. "I have to say my small grandson was most disappointed by this design which he thought was very poor spirited," she laughs. "He wanted us all to swing across by a rope!"

Meanwhile, downstairs there was a very small kitchen to deal with. "It was narrow, like a corridor. But by blocking off a door and moving a wall by around 2m [reducing the dining room to 11m in length], it gave me a much bigger space – big enough to be used as a breakfast room."

"Although the initial building work was done efficiently and quickly, it has taken two and a half years to complete all the interior finishing off and decorating. It would be hard to find another house with such space, surrounded by unspoilt countryside yet three minutes drive from a large supermarket. I even have a view of the sea – if I stand on the garden table!" ■

FLOORPLAN

All the major structural amendments were to the ground floor of the property. The large dining room and a remodelled kitchen space are the main features.

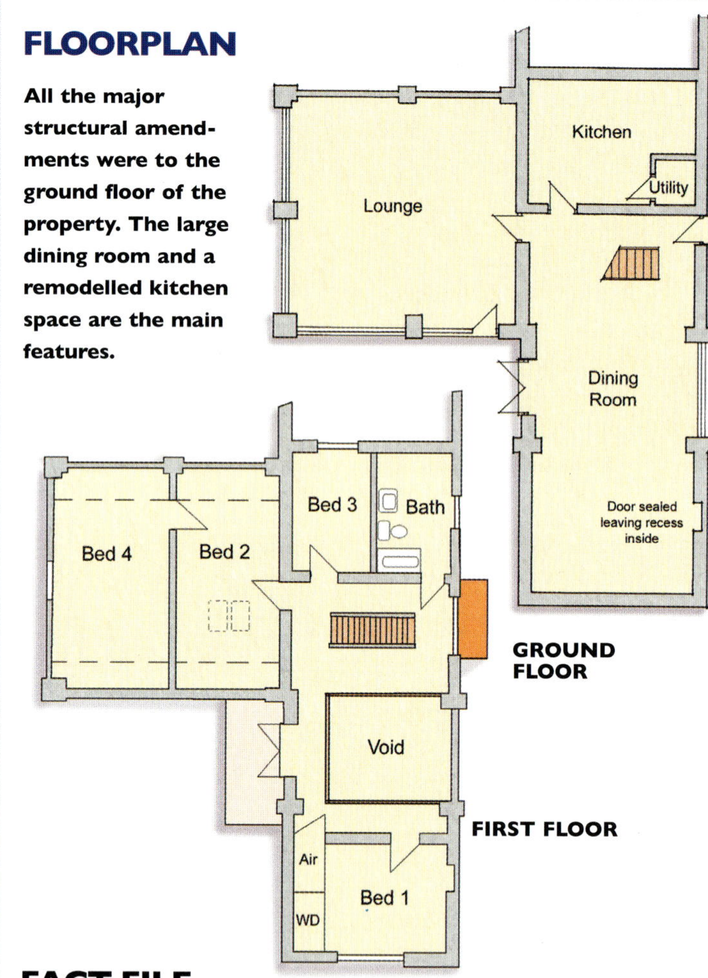

FACT FILE

Costs as of Nov 2003
Name: Gillian Ashby
Profession: Full time housewife and voluntary worker
Area: Devon
House type: Four bedroom Victorian barn conversion
House size: Approx 258m²
Build route: Main contractor
Construction: Brick with timber upper section at rear and tiled roof
Finance: Private
Build time: Six weeks
Property cost: £138,000
Build cost: £12,000
Total cost: £150,000
House value: £290,000
Cost/m²: £600

48%
COST SAVING

USEFUL CONTACTS

Builder – Alan Blackie of Sidmouth Window Company: 01395 579999; **Kitchen** – Howdens Joinery Ltd: 01392 413446; **Damp proofing** – Terminex Property Services: 01752 840600; **Lighting** – John Hill's Lighting: 01392 276371

A NEW HOME BUILT FROM AN OLD BARN

THE FINAL FRAME

Retired housebuilder Brian Barnett's final project was a new home for himself and wife Loretta, created using an oak frame reclaimed from an 18th Century barn.

WORDS: MICHAEL HOLMES PHOTOGRAPHY: NIGEL RIGDEN

OLD TIMBER BARNS tend to come with lots of potential and plenty of character. Unfortunately – as new owners very quickly find out – they also tend to come with lots of design constraints. The siting and orientation are a fait accompli; the planners usually want to see as few alterations to the original structure as possible, limiting the number of new window and door openings; then there is the cost of working with an old structure in situ – far greater than building from scratch thanks to underpinning, damp proofing, timber treatment, retro fitting insulation and so on.

Given its sensitive green belt location near Kingswood, Surrey, and close proximity to a Grade II listed farmhouse, you would expect Brian and Loretta Barnett to have experienced all manner of problems when they turned their mid 18th Century oak barn into a beautiful three bedroom home – but then theirs is no ordinary barn conversion. Although the frame is well over a century old, it only arrived in its current location three years ago and is technically, therefore, a new house and as such free of the usual planning constraints.

Brian and Loretta purchased the barn from Peter Barker of Antique Buildings Ltd, a firm which specialises in reclaimed oak and elm frames salvaged from all over the British Isles. At his yard in Dunsfold, Surrey, Peter has a selection of frames of all shapes and sizes, ranging from old cowsheds, to hovels and barns. The Barnetts' frame, which cost them £16,000, originally housed cattle on a farm near Lingfield, around twelve miles from Kingswood. "When we first set eyes on it, it was nothing more than a pile of timber accompanied by a few photographs and some measurements," recalls Brian.

"The outline planning permission on our site had only been granted on appeal, as a replacement for an existing rundown cottage.

The chimney and brick inglenook were all built using local bricks from Chelwood Brick.

"IT MADE SENSE TO TURN THE ACCOMMODATION UPSIDE DOWN, WITH THE FIRST FLOOR GIVEN OVER TO ONE LARGE LIVING AREA."

The ground floor is covered in Chinese slate and the first floor in ceramic quarry tiles. Both are warmed by a Nu-Heat underfloor heating system.

The windows were all made from new oak. Great care was taken to line up the frames with timbers of the barn.

Given the Green Belt location, one of the planning conditions was that the new dwelling be no more than ten percent larger in volume than the original. The frame we bought, measuring 12m x 6m plus an outshot, was more or less a perfect match for what we were allowed."

With the exception of the sole plates, which had been in contact with the ground, the frame was in good condition. These were replaced in their entirety using 8" x 6" sections of green oak. "We also had to replace the lean to section of the barn to create the third bedroom and lean to overhang for the log store," explains Brian. "I would say that you have to allow a figure of around twenty percent on top of the price of the frame for replacement oak. None the less, this route has lots of advantages compared to converting an existing barn. It would have been more expensive to have worked in situ.

"With a project like this, it is essential to find someone who knows

"I WOULD SAY THAT YOU HAVE TO ALLOW A FIGURE OF AROUND TWENTY PERCENT ON TOP OF THE PRICE OF THE FRAME FOR REPLACEMENT OAK."

about barn construction," Brian advises. "We phoned several firms and were lucky enough to find an experienced master carpenter, Gary Wood, who took great pains to get things right and really was first class."

The Barnetts' frame was first rebuilt in sections laid flat out on the ground to ensure that all of the parts were complete and the joints in working order. Any missing or damaged timbers were repaired or replaced using either reclaimed oak or new green oak which Gary hand ➤

"MY ADVICE FOR ANYONE TRYING TO UNDERTAKE A PROJECT LIKE THIS IS NOT TO TRY AND DO IT ON A SHOESTRING."

The irregular size and shape of the clay tiles – a combination of three different colours from Tudor Tile Co – lend the building considerable character, helping it to look established in its setting.

adzed to give an aged effect. The whole structure was then treated, before being dismantled and erected once more, stick by stick, with each joint held together using freshly cut oak pegs. Three years on, it is very difficult to distinguish repair from original, or old from new.

"Given the restricted volume we had to work with, it would have been impossible to create double height spaces and still make the most of the timber roof structure. It therefore made sense to turn the accommodation upside down, with the first floor given over to one large open plan living area, making the most of the barn roof and the views over the surrounding countryside," explains Brian. "Once we had made this decision, the rest of the house designed itself around the shape of the frame."

In spite of his prior experience, Brian decided to bring on board an architect to fine tune the design and handle the planning process. "You can get too close to these things," he says, "plus my architect, Norman Franklin, is well known to the local planners who trust his work. His reputation inevitably helped in the negotiations. I also brought on board a structural engineer to prove that the frame was up to the job."

In order to create the ceiling heights necessary for two storeys, the frame was erected on top of a 1200mm blockwork plinth which is clad in knapped

The simple, rustic style painted kitchen units were made by local carpenter David How for half the cost of a designer brand name kitchen.

flint panels with brick surrounds from the local Chelwood brickworks. The exterior of the oak frame is clad in large section feather edged boarding, stained black.

"One feature which the planners did object to was the chimney – which they claimed was not in keeping with a barn," says Brian. "However, Norman managed to persuade them that the design would be sympathetic to the character of the building and subtly reminded them that this was, after all, a new house and not a barn conversion."

"The original budget for the build was £100,000, but we ended up spending £150,000," says Brian. "Much of the extra cost went on the quality of the materials, such as the Tudor Handmade clay tiles, flint panels and so on, but they are worth every penny.

"My advice for anyone trying to undertake a project like this is not to try and do it on a shoestring," he advises. "The building dictates to you and you can't cut corners – it would stand out like a sore thumb if, for instance, you put in a standard softwood staircase instead of oak. The only downside to a house like this is that there is nowhere to hang pictures in amongst all of the beams, but then they are a feature in their own right!" ■

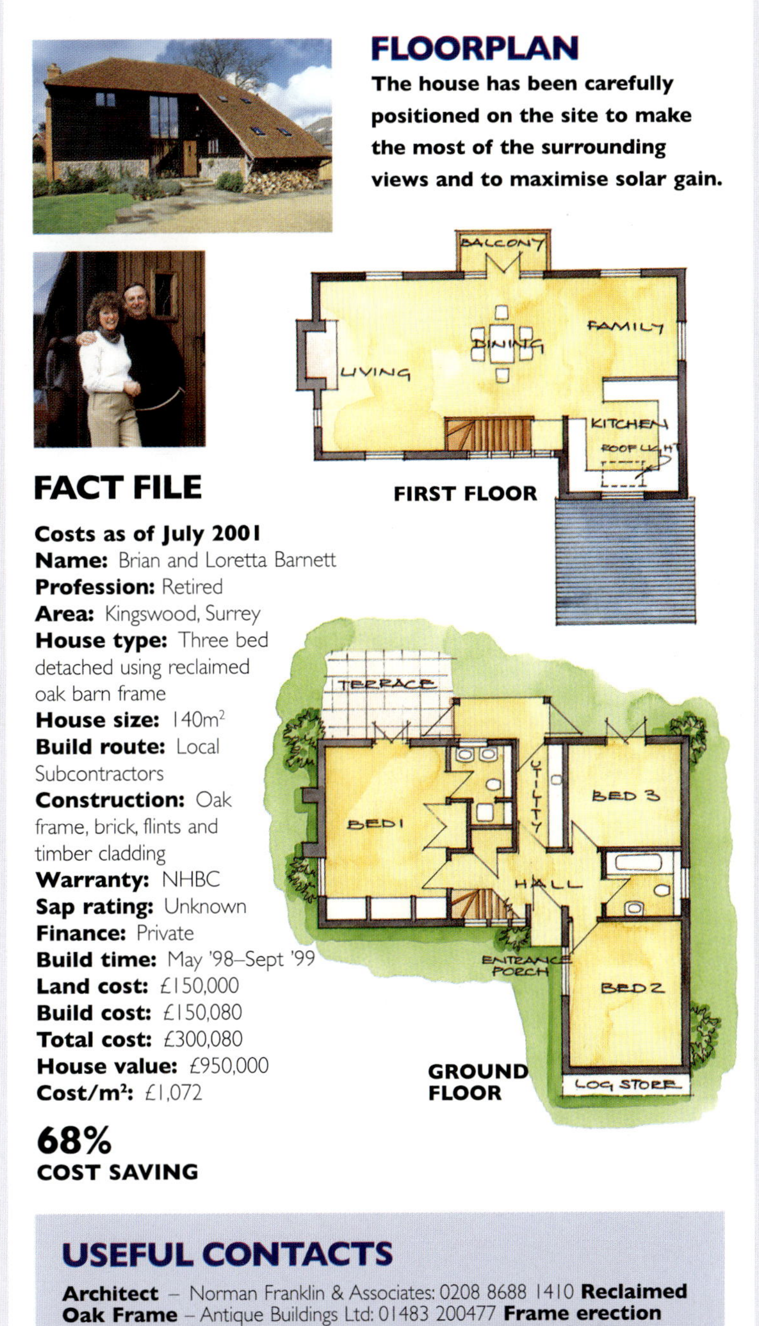

FLOORPLAN

The house has been carefully positioned on the site to make the most of the surrounding views and to maximise solar gain.

FACT FILE

Costs as of July 2001
Name: Brian and Loretta Barnett
Profession: Retired
Area: Kingswood, Surrey
House type: Three bed detached using reclaimed oak barn frame
House size: 140m²
Build route: Local Subcontractors
Construction: Oak frame, brick, flints and timber cladding
Warranty: NHBC
Sap rating: Unknown
Finance: Private
Build time: May '98–Sept '99
Land cost: £150,000
Build cost: £150,080
Total cost: £300,080
House value: £950,000
Cost/m²: £1,072

68%
COST SAVING

USEFUL CONTACTS

Architect – Norman Franklin & Associates: 0208 8688 1410 **Reclaimed Oak Frame** – Antique Buildings Ltd: 01483 200477 **Frame erection and restoration** – Gary Wood: 01737 842533 **Bricks** – Chelwood: 0161 485 8211 **Rooftiles** – Tudor Roof Tile Co Ltd: 01797 320202 **Underfloor Heating** – Nu-Heat Ltd: 01395 578482 **Conservation Rooflights** – The Velux Co: 01592 772211 **Oak Doors** – British Gates & Timber: 01580 291555 **Woodburning Stove** – Dorking Stove Co: 01306 883201 **Kitchen** – David How: 01306 611092 **Range Cooker** – Britannia: 01253 471111

The listed barn was originally part of the Beau Manor estate in Leicestershire and was subject to strict Conservation Area planning requirements.

CONVERTING A BARN FOR UNDER £100,000

COUNTRY CHARACTER

Christine Smeaton has created an elegant home with a relaxed country feel by converting an old barn.

WORDS: DEBBIE JEFFERY PHOTOGRAPHY: JEREMY PHILLIPS

I AM NO good with money," admits retired occupational therapist Christine Smeaton, "but when it came to converting my barn I had to watch every penny." Following a change in her personal circumstances Christine had to leave Rushall Fields Farm, the 17th century family home where she had lived for almost 30 years. "I didn't want to leave, but I decided that I would retain the barn we had used for keeping hay, chickens and horses. I had considered converting it in the past and felt that it would make a pleasant home – although my friends, family and accountant tried hard to dissuade me!"

Time and again Christine was warned that she may feel isolated in such a rural location, and that conversion projects can sometimes end in financial disaster. For her, overspending was not an option. "I had just £90,000 with no room for error," she explains. "Even so, I was adamant that I wanted to tackle the conversion, and no amount of opposition could change my mind. I needed a home far more than foreign holidays or new clothes, and sunk every penny into the venture."

Christine vacated the Grade II listed farmhouse and considered buying a caravan. In the end, however, she moved into rented accommodation near Loughborough and decided to make no major decisions for the next six months. After this time she approached a local architect to design her new home. Initially, she planned to base an occupational therapy centre for the elderly at the barn, with provision ➤

made for a reception, therapy lounge and a new therapy workshop, built on the footprint of an existing brick and corrugated sheet lean-to.

"It became apparent that the design, which positioned my living room and bedroom upstairs, was not going to be affordable," says Christine, who admits to having had some doubts at this stage.

Fortunately Peter Watson, who had purchased Rushall Fields Farm, is an NHBC registered builder who saw Christine's plight and offered to help. Concerned that she might sell the barn, he proposed to carry out the conversion on her behalf for a fixed quote.

"Peter explained to me that, if we kept to the basics, we should be able to do it," says Christine, who was impressed when she visited another barn which Peter had converted. "It would be the extras that would push me over budget, and so he drew up a detailed quote, itemising provisional sums for everything from the services to the light fittings and decoration."

Included in the quote were drawings and plans with specifications for building regulations. The grass verge from the barn to the water main would be excavated to connect the barn to mains water and, within the same trench, an electrical service duct was laid. The new foundations required to stabilise the building were assumed to be a maximum of one metre deep from the existing ground level, and Peter explained that any additional work would be negotiated and charged for accordingly. Christine decided to

Keeping a tight control over the budget allowed Christine to spend money on elegant details such as the Clayton Munroe door furniture. ➤

"I LOVED THE IDEA OF AN UNFITTED KITCHEN, WHICH I FELT WOULD PERFECTLY COMPLEMENT THE RELAXED NATURE OF THE BARN..."

accept the quote and planning was ultimately granted for the conversion.

The new design incorporates a kitchen, utility room, cloakroom, open plan lounge and dining room on the ground floor, with two bedrooms and a bathroom upstairs. One lean-to extension would be demolished, leaving the 230m^2 barn and single storey utility room to the east elevation, with a large, full height window to the west, where the original barn doors had been. Although Christine changed her mind several times regarding the barn's layout, she eventually settled on this simple floor plan for both financial and practical reasons. "It was straightforward and unfussy – just what I needed," she explains.

Set in the beautiful village of Old Woodhouse, approximately four miles south of Loughborough, the listed 18th century barn is situated in a Conservation Area. This dictated that the original random stone walls and soft red brick corner jambs, heads and cills be retained and that any new window openings had to be kept to a minimum. Part of the building was roofed in corrugated sheeting, which was to be replaced by second-hand Welsh slates, whilst a collapsing gable end wall needed to be reconstructed on new foundations to match the existing building, with a damp proof course at ground floor level.

The old concrete yard was broken up and re-used as a sub-base for access and parking areas, with the building excavated internally prior to the construction of an insulated floor slab. Incredibly, the massive new septic tank was stolen from the site before it could be installed and man-handled across neighbouring fields by cheeky thieves!

Reclaimed pine internal doors and exposed beams give the barn a rustic feel.

"SITUATED IN A CONSERVATION AREA... ANY NEW WINDOW OPENINGS HAD TO BE KEPT TO A MINIMUM."

External walls were repaired and repointed, and any badly perished bricks chopped out and replaced with matching facing bricks. Materials were cleaned and re-used wherever possible in order to keep costs down, and Christine visited the site every day to ascertain that the project was on target and within budget.

"I had explained to Peter that I can be quite demanding to work with," she states, "but he listened patiently to my requests and told me exactly what I needed to choose for each stage of the five month build. I like to see things in person rather than as a picture in a catalogue, so this entailed a fair amount of travelling around towards the end. Peter worked so fast that I sometimes struggled to keep up with his requests for all the fixtures and fittings I needed."

Amazingly, considering her tight budget, Christine's numerous shopping trips paid off. She managed to save money on certain things, including the sanitaryware, which was then spent on items such as the Clayton Munroe door furniture and reclaimed pine doors. These details make all the difference to the finished conversion, taking it out of the ordinary and giving the impression of a far more expensive project.

The kitchen is the most eye-catching room in the house, with its bespoke handcrafted Devol kitchen units and dresser. "I loved the idea of an unfitted kitchen, which I felt would perfectly complement the relaxed nature of the barn," says Christine.

"Initially, I enquired about purchasing just one unit, because I didn't think I could afford anything else!" Devol's proprietor and designer, Paul O'Leary visited the barn and, working within Christine's budget of £5,000, managed to create a simple working room full of charm.

"Many kitchen companies offer a free design service but I would question its value," says Paul. "The salesman's goal is clearly to squeeze in as many bought-in units as will fit. If anything, I try to advise clients that less is more. Kitchens, especially small ones, need space as well as storage."

Christine already had an oven, but Paul convinced her to invest in a new stainless steel Smeg range style cooker to complement the hand painted kitchen furniture. She also ordered a butcher's block and an open fronted dresser which provides additional storage for china.

"When it comes to choosing colours I have absolutely no idea," laughs Christine, who enlisted her daughter's advice on decoration. "I would probably have ended up painting the barn red, white and blue! Paul helped me design the entire kitchen, including items such as the Portuguese floor tiles, and people are quite surprised when they see how restrained and tasteful it all is – they wonder how I managed it!"

After 15 months living in rented accommodation Christine Smeaton was able to move into her new home the day after the decorators moved out. "It is exactly what I wanted – there's nothing I would change," she enthuses. "My family wondered how I would cope living next door to my old home, but it doesn't bother me at all. I have moved on to the next stage in my life, and converting the barn has given me financial freedom and a beautiful home. I could never have afforded to buy such a lovely property in this area. It all worked out perfectly." ■

FACT FILE

Name: Christine Smeaton
Profession: Retired occupational therapist
Area: Leicestershire
House type: Two bedroom barn conversion
House size: 230m²
Build route: Building contractor
Construction: Stone and brick walls, slate roof
Finance: Private
Build time: April '02 - Aug '02
Land cost: Already owned, valued at £60,000 without planning permission

Build cost: £100,000
Total cost: £161,000
House value: £350,000
Cost/m²: £465

54%
COST SAVING

Cost Breakdown:

Mains electricity to site	£2,700
Mains water to site	£1,000
Building regulation fees	£800
Structural engineer	£200
Damp proofing and timber treatment	£1,200
Construction and materials	£75,330
Plumbing and heating	£6,200
Electrics and kitchen lighting	£1,800
Kitchen and appliances	£4,700
Sanitaryware	£800
Light fittings	£200
Paint and decoration	£3000
Wall tiling	£500
Doors, windows and glazing	£6,800
Staircase and balustrading	£1,500
VAT Refund	(£6,800)
TOTAL	**£99,970**

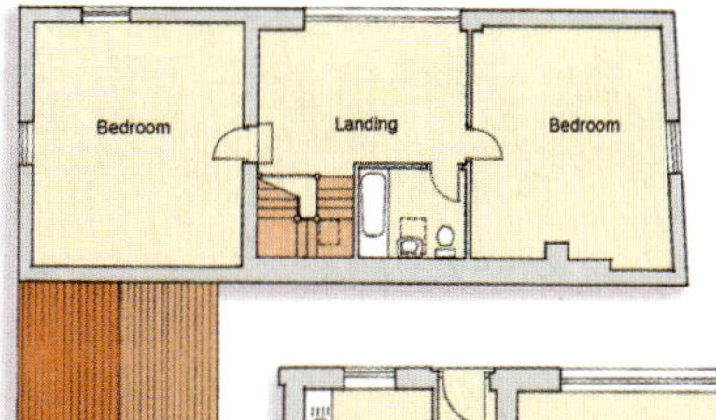

FIRST FLOOR

FLOORPLAN

GROUND FLOOR

Kitchen
Dining
Living Room
Boiler
Utility

The barn has a simple layout, with two bedrooms and a bathroom on the first floor and a ground floor kitchen leading into the open plan living and dining room.

USEFUL CONTACTS

Architect – Logos Designs: 01509 621001; **Contractor** – Peter B. Watson Ltd.: 01509 890688; **Reclaimed doors** – Quorn Pine: 01509 416031; **Door furniture** – Clayton Munroe: 01803 865700; **Weathervane restoration** – Original Forgery: 01398 331400; **Floor tiles** - Homebase: 0845 077 8888; **Garden design** - Elizabeth Younger: 0131 3326359

The walls are of local rubblestone at the front and some local brick at the rear and the roof tiles are clay double Roman pantiles.

RESCUE PACKAGE

George and Liz Robertson have created a beautiful new home by renovating a 150 year old barn that had been poorly converted in the 1970s.

WORDS: VICTORIA JENKINS
PHOTOGRAPHY: HUGH BURDEN

WHEN GEORGE AND Liz Robertson found The Old Dairy in Somerset it had already been converted from a 150 year old barn into living accommodation. But there was still a lot more work to be done – nearly £130,000 worth in fact – and it took a full 18 months to complete.

"I thought the house was fine but Liz was making mental notes to herself about what needed to be changed," says George. "I can now see how much better it looks – altogether more light and airy. For instance, where we now have oak floors where there was concrete screed with red carpet before. It was rather depressing."

The first thing the couple did was to call in architect Trevor Spurway, who realised he had two objectives. One was to give the first floor a more traditional appearance – "with its low Artexed 1970s ceilings it looked rather like a flat," says George – and the other was to create a home with a layout that would match the couple's informal lifestyle.

"This included adding a big extension – measuring some 26m² – onto the kitchen to create an open plan dining area, plus a covered outdoor area for eating and barbecuing," says Trevor. "The idea was to integrate the ➤

The kitchen changed shape when the Garden Room extension was built. The floor is of Spanish limestone and the radiator is an 'Iguana' by Belgian manufacturer Jaga.

garden with the dwelling so there was a cross-over. If you were in the extension you felt you were in the garden and vice versa."

Phase one of the works began with the first floor, which took three months to complete. Here the ceilings were either raised or removed to expose the traditional barn trusses for a lofty barn-like feeling. This is particularly striking in the master bedroom. The original bathroom was converted to create a dressing room and the study/ third bedroom was converted to create a new en suite bathroom. The original second bathroom was refurbished and is now a shower room.

Phase two of the project involved more major building work and took six months. The main job was to build a substantial extension to add extra living space. This area has tall open raftered ceilings to create a barn-like effect and give a greater feeling of space. "To ensure the extension fits in with the existing barn we used traditional materials. The walls were finished in render with brick quoins and windows surrounds, and we used reclaimed double Roman clay tiles on the roof," says Trevor.

The extension, known as the Garden Room, adjoins the kitchen – which itself had to be newly fitted out, despite not being in bad condition, to better suit the new open plan layout. They called in a Bridgwater interior design supplies firm called Original Building Supply Company who designed and built them a bespoke kitchen with new oak Shaker style units and black granite worktops. They also supplied the Miele electric oven (incorporating a steam oven and plate warmer), a Miele induction hob and Miele hood. ➤

The sitting room adjoins the Garden Room, which has an American oak floor and is some 30m2.

"THE NEW BOILER POWERS A MAINS PRESSURE 'UNVENTED' HOT WATER SYSTEM TO GIVE SUFFICIENT FLOW FOR THE MODERN TAPS THROUGHOUT THE HOUSE."

"Our oven combines certain advantages of both conventional and microwave ovens as it utilises large volumes of steam, delivered to the cavity at a preset temperature, in order to cook the food without direct immersion in water – particularly good for vegetables and fish," says George.

"We don't have mains gas but that is not the reason we chose an induction hob," he adds. "We prefer induction heating for its extremely rapid and responsive nature (in that sense it is a bit like gas), and also because the hob itself does not generate any heat – it is generated in the vessel itself, through induction. Also, food residues do not burn to its surface and it is very easy to clean. Induction hobs, which are not really more expensive than conventional ceramic hobs, are uncommon in the UK."

The new boiler runs on oil and powers a radiator central heating system and a mains pressure 'unvented' hot water system – necessary to give sufficient flow for the modern taps utilised throughout house. Limestone flooring from Spain was laid in both kitchen and garden room and new American oak wide board flooring was laid throughout the rest of the downstairs level.

Outside the garden room, a slate patio and an oak canopy were installed and a new front door with a porch was built further along the centre of the house. This is because the original front door was in the style

The master bedroom has the original 150 year old barn trusses exposed.

of a narrow portcullis – so narrow in fact that the previous owners gave up trying to get their oak dining table through it when moving out. "So they left it behind in the hallway. We've got it now!" says George.

The original white powder-coated aluminium window frames in hardwood sub frames were not appropriate for the style of the building and so were replaced with new hardwood window frames with a stained finish.

Another small project was the renovation of a small former barn store which has now become a TV room. "When we found it, it was just a long open room painted orange with carpet tiles on the floor and a woodburning stove. The windows were falling apart and the door was shabby," says George. The door and windows have now been replaced with new ones of hardwood and a new laminate floor and radiators have been installed to make it cosy.

"The extension, patio, porch and windows cost £86,000 altogether," says George. "The windows alone cost £18,000, as there are 24 of them in all."

The Robertsons haven't finished yet, however. "We now have only two bedrooms and a dressing room," says George. "And we would like five altogether, plus another bathroom. But as we have an enormous garage attached to the house we now plan to have it converted into the extra bedrooms and bathroom – and then we'll have a new detached garage built in its place." ■

FACT FILE

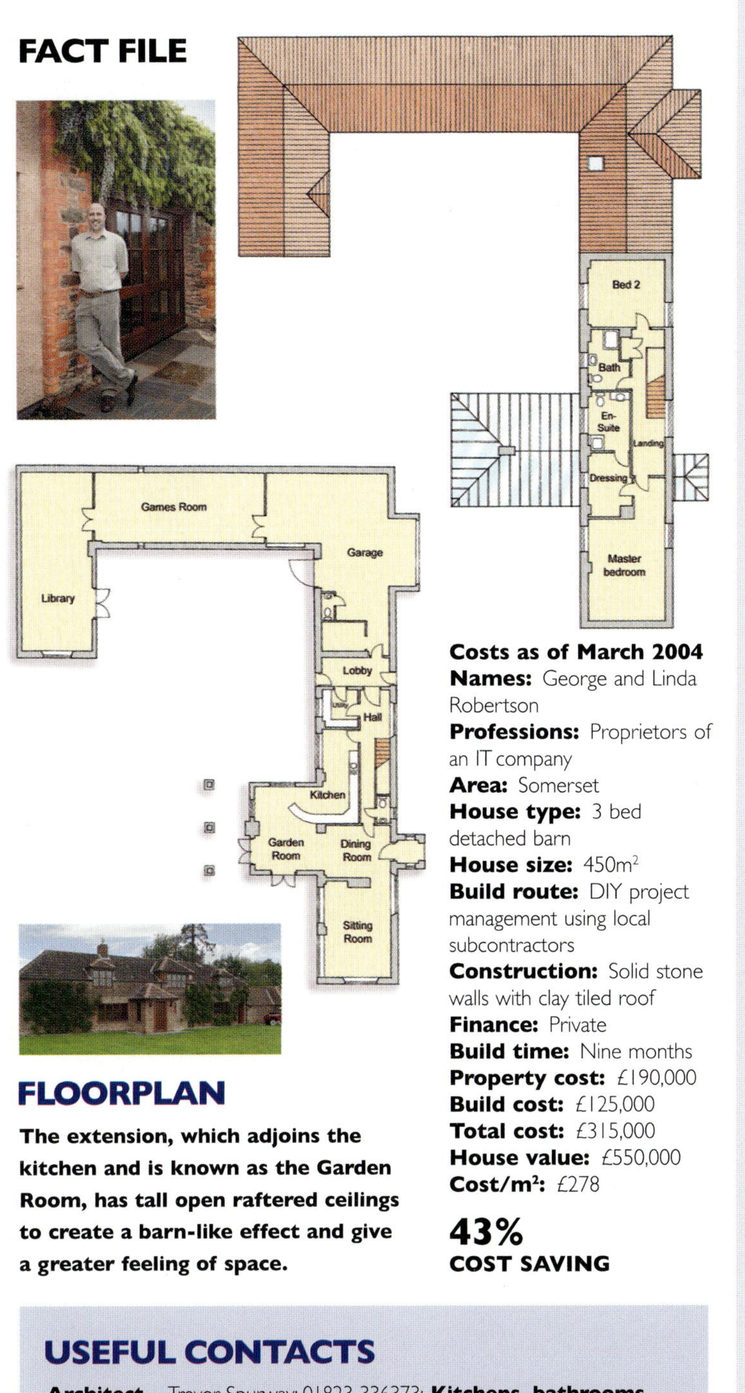

Costs as of March 2004
Names: George and Linda Robertson
Professions: Proprietors of an IT company
Area: Somerset
House type: 3 bed detached barn
House size: 450m^2
Build route: DIY project management using local subcontractors
Construction: Solid stone walls with clay tiled roof
Finance: Private
Build time: Nine months
Property cost: £190,000
Build cost: £125,000
Total cost: £315,000
House value: £550,000
Cost/m^2: £278

43%
COST SAVING

FLOORPLAN

The extension, which adjoins the kitchen and is known as the Garden Room, has tall open raftered ceilings to create a barn-like effect and give a greater feeling of space.

USEFUL CONTACTS

Architect – Trevor Spurway: 01823 336373; **Kitchens, bathrooms and other interior design supplies** – Original Building Supply Company: 0870 0110 808; **Builder –** Adsborough Builders of Taunton: 01823 332132; **Bathroom** – Ripples: 0117 97311 44

INSPIRATION IN OAK

Blair and Liz Nimmo's spectacular 910m^2 new oak framed home has been built without the usual constraints of size or budget.

WORDS: DEBBIE JEFFERY PHOTOGRAPHY: JEREMY PHILLIPS

The dining hall (opposite page) stands at the heart of the house, where the stunning oak frame forms galleried landings on the first and second floors and rooflights create the feel of an atrium.

LOWER HOUSE FARM is the kind of property most of us can only ever dream of owning. The 18-acre estate is set in one of the prettiest parts of Cheshire and enjoys uninterrupted panoramic views. In addition to the main house, outbuildings and landscaped grounds, there is an indoor pool complex, complete with sauna and steam rooms. An oak framed pavilion stands beside the lake and there are orchards, woodland and a wild flower meadow.

Chosen for its privacy and magnificent views, the site offered Liz and Blair Nimmo the opportunity to replace the existing farmhouse with an altogether more ambitious and architecturally interesting building. The couple have four children – Lucy, Tom, Ben and Beth, aged between 17 and 21 – and had been searching the area for a significant period house without success.

"We were living in rented accommodation and soon realised that there were very few interesting rural properties available in this part of Cheshire. Those we did discover were sold privately which made it virtually impossible for us to get what we wanted," explains Blair. "There was no other choice – we had to build."

Blair immediately began sketching ideas for his dream house, but Liz suggested that it would be interesting to verbally agree the kind of accommodation they needed and come up with a design in words. The resulting synopsis was: 'a family home with a few large rooms, no corridors and lots of natural light'.

The kitchen has been designed as the focus for family life and the raised aisle barn format of oak framing gives an enormous sense of space. Subtle lighting behind the purlins illuminates the sarking boards and dimmable halogens light the dining and preparation tables.

"ALTHOUGH WE HAD ALREADY DESIGNED A LAYOUT FOR THE HOUSE, WE HAD NO IDEA ABOUT ARCHITECTURAL STYLE. WE DIDN'T WANT A MOCK PERIOD HOUSE OR SOMETHING TOO CONTEMPORARY." ➤

"WE SPENT A HOLIDAY IN MOROCCO, WHERE HOUSES ARE OFTEN BUILT AROUND COURTYARDS FOR SHADE AND PRIVACY, AND THIS INSPIRED THE CRUCIFORM SHAPE OF OUR OWN HOUSE."

"We wanted to get away from the idea of having numerous rooms which are never used, and to build a house that would fit the way we live, with much of family life revolving around the kitchen," says Liz. "We spent a holiday in Morocco, where houses are often built around courtyards for shade and privacy, and this inspired the cruciform shape of our own house, which allows each wing to benefit from triple aspect views."

The Nimmos disliked the idea of a formal dining room, and instead the three storey house has been designed with a dramatic central dining hall where double doors lead to each of the four wings, the oak frame forms galleries on the first and second floors and rooflights create the feel of an atrium.

Leading off from this hallway is a spacious 'living' kitchen, overlooked by a gallery study, with a large stone fireplace providing the focal point at the far end of the room. Blair describes the kitchen as "a house within a house," where cooking, eating and sitting areas are combined and two sets of French windows open out onto a stone flagged veranda, which wraps around the kitchen wing and provides access to the kitchen garden on one side and a large terrace and outside dining area on the other.

On the first and second floors the wings house five separate suites, all similarly designed with a bedroom, walk-in wardrobe, wet room and a sitting room/study. There are even secret storage areas with cleverly concealed doors. "With four children we knew that we needed space for visiting friends, but disliked the idea of designated guest bedrooms which would be empty for much of the time," says Liz. "By creating individual areas for the children they have enough space to entertain friends in their own rooms, and it means that everyone has their own quiet retreat."

Finding a suitable site for their design proved far easier than the Nimmos had imagined, and they soon discovered Lower House Farm and were charmed by its location and views. The dilapidated farm range was particularly pretty, and has been repaired and retained as

The vaulted kitchen and living area is the place where family life revolves. ➤

The drawing room has been designed as a home cinema and has a secret staircase behind the fireplace which connects with the master bedroom above.

additional guest/leisure accommodation, but the main farmhouse was a bland brick structure. Architecturally, the property had no redeeming features, and was ripe for redevelopment.

"Although we had already designed a layout for the house, we had no idea about architectural style," says Blair. "We didn't want a mock period house or something too contemporary so, when a friend showed us an article about Roderick James' oak framed buildings, we knew we had found the answer."

Liz and Blair worked with Roderick to adapt their design before appointing Jim Brotherhood as their locally-based project architect. Jim developed the plans and achieved consent for the main house, swimming pool and garages, while the oak frame details for the house and pool were designed by Duncan Ellis, who sadly died in 2001. "He was an inspired engineer and great fun to work with," says Blair, who dedicated the pool to his memory. "We would email ideas and comments to one another at 3am."

One of the first tasks was to make a 'site office' in the form of an idyllic oak framed lakeside pavilion with oak shingles, complete with a boat jetty cantilevered out over the water. Built as a birthday gift for Liz, the pavilion

has hot and cold water, a telephone and wood burner, and it was here that Blair spent much of his time during the build, which began in April 2000 with the demolition of the main house. An existing cellar was discovered, and has been retained – with the new four layer construction incorporating external drainage, blockwork, asphalt tanking and facing brickwork.

"A quantity surveyor friend suggested a building contractor and offered to oversee the project for us, proposing a management contract which involved tendering for quotes for every trade at each stage of the build," Blair explains. "We parcelled the work up as the detailed drawings were completed which enabled us to start on site before everything was finalised. I found the whole building process absolutely fascinating, and felt that I was getting extra value by being here and seeing how things were done."

Blair is an electronic engineer and self-confessed perfectionist, and great attention to detail has been paid to fixtures and fittings throughout the house. There is a dedicated communications room with Cat-5 wiring for a structured network operating from a central hub, a comprehensive intruder and fire alarm system, ISDN digital telephone lines, CCTV monitoring and satellite broadband. Opening skirting boards, based on a design Blair devised at the age of 13, conceal sockets and wiring from view. ➤

"WE WENT A BIT OVER THE TOP WITH THE POOL, BUT THE REFLECTION OF THE OAK BEAMS IN THE WATER IS REALLY QUITE INCREDIBLE."

Robust natural materials have been used wherever possible: there are slate and end-grain beech surfaces on the handcrafted kitchen units, and the floors are a mixture of solid elm, oak, pitch pine, limestone and handmade terracotta tiles. A sophisticated underfloor heating system has been laid and the property is highly insulated, with Pilkington double glazed 'K' glass in oak framed panels – contrasting with the single glazed handmade cylinder glass which has been used in the brick wings.

The swimming pool complex is a self-contained green oak structure incorporating hammer beam truss oak frames. The pool has a level water deck and underwater lighting, there are balconies to each end, an air management system designed to eliminate condensation and heat and waterproof speakers in the sauna and steam room. "We went a bit over the top with the pool," admits Blair, who swims every day, "but the reflection of the oak beams in the water is really quite incredible, and the full-height glazing means that you can swim and enjoy the view at the same time.

The swimming pool complex is a self-contained building with Indian stone flooring, a sauna and steam room.

"The house has proved ideal for parties and family gatherings, and we have thoroughly enjoyed living here but, now that the children are older, we have decided to sell Lower House Farm in order to design and build a smaller house beside the water. I love to sail, and we want to lure our children and grandchildren to visit us for seaside holidays, so we will once again be looking forward to working with Roderick James to build an oak framed home for the next stage of our lives." ■

FLOORPLAN

The three storey house revolves around a central dining hall, with doors to the kitchen wing, the drawing room, study, snug and service areas. Five bedroom suites incorporate wet rooms and living/study areas, with access to the balcony on the first floor.

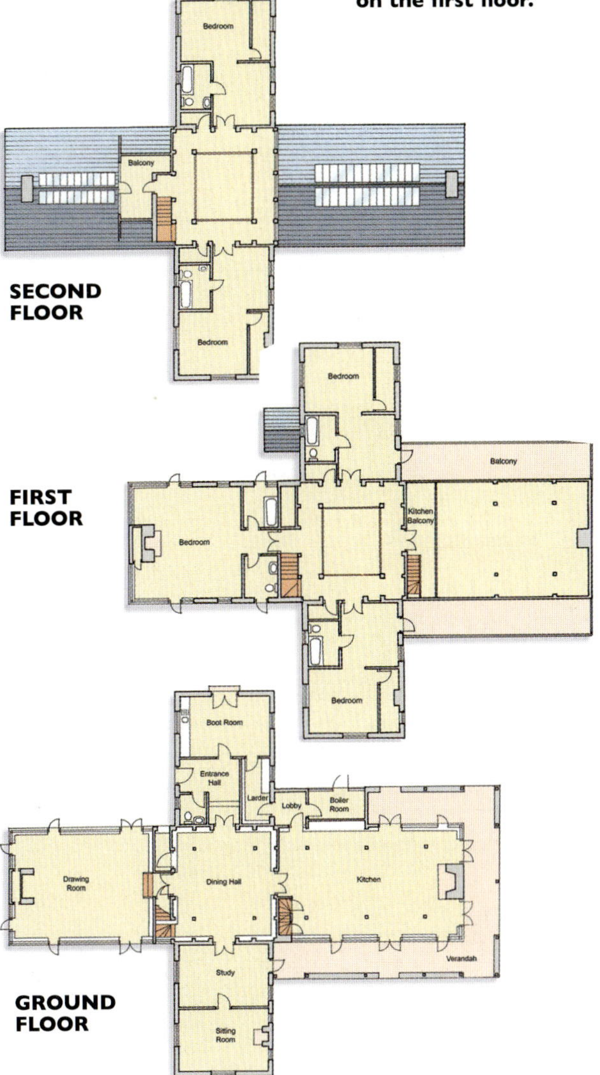

FACT FILE

Names: Liz and Blair Nimmo
Professions: Homemaker and owner/MD of fibre optic company
Area: Cheshire
House type: Five bedroom detached house with separate pool complex and outbuildings
House size: 905m^2
Build route: Main contractors and subcontractors
Construction: Brick and block, oak framing, clay roof tiles
Warranty: Architect's Certificate
Finance: NatWest mortgage
Build time: April '00 – Aug '01
Land cost: £702,000
Build cost: £2,010,000
Total cost: £2,712,000
House value: £4.5m
Cost/m²: £2,220

40%
COST SAVING

Cost Breakdown:
(not including pool, farm range, pavilion etc.)

Architect's fees	£220,000
Main contractor management fees (all phases)	£310,000
Groundworks, foundations, slab and plinth wall	£116,000
Bricklaying	£120,000
Oak frames, erection and sandblasting	£317,000
Joinery	£399,000
Roofing, tiling, lead and balcony	£64,000
Plastering	£31,000
Flooring	£93,000
Cellar tanking	£4,000
Plumbing	£47,000
Electrics	£55,000
Painting and decoration	£13,000
Fireplaces	£8,000
General building work	£6,000
Services (incl. drainage)	£27,000
Landscaping and planting	£60,000
External works	£120,000
TOTAL	**£2,010,00**

BUILDING IN OAK

Post and beam oak framing techniques have remained largely unchanged for centuries, yet this most traditional construction method lends itself equally well to contemporary style buildings. The key to its success is in the barn-like vaulted double-height spaces and the prominence given to the structure itself. For a full list of oak frame suppliers visit www.homebuilding.co.uk and search on 'oak frame'.

USEFUL CONTACTS

Style architect – Roderick James Architects LLP: 01803 732900; **Project architect** – James Brotherhood: 01244 579000; **Oak framing –** Carpenter Oak Ltd: 01803 722900; **Construction management** – Hantall Developments: 0161 368 5885; **Reclaimed and hard materials, York stone, Indian stone, sandstone wall blocks, roof tiles, bricks, sandstone setts** – Blackham Reclamation: 01948 820658; **Reclaimed and new timber (oak, elm, pitch pine etc)** – Edwards of Cheshire: 01925 631678; **Swimming pool design and construction** – Clearwater Swimming Pools: 01865 766112; **Design and supply of sauna, steam room etc** – Dream Leisure: 020 8977 9900; **High quality joinery and construction management** – Lodge Joinery: 07931 556517; **Surfacing (tarmac, farm tracks, chipped finishes etc.)** – Miles Macadam: 01948 820489; **Stone laying (slabs, paths, setts)** – Pave4u: 07753 826913; **Interior design** – Interiors with Briony Cooper: 01743 235702; **Sanitaryware** – Villeroy & Boch: 020 8871 4028; **Paints** – Zoffany: 0870 830 0350; Farrow & Ball: 01202 876141; **Taps** – Vola (UK) Ltd: 01525 841155

So simple – It's **brilliant!**
Monodraught
Sunpipe
The science of channelling free natural daylight
to places that windows cannot reach.
For a more illuminating insight to our unique product range...
t : 01494 897700 w : www.monodraught.com